T0271043

Screw Theory and Its Application to Spatial Robot Manipulators

Discover a fresh take on classical screw theory and understand the geometry embedded within robots and mechanisms with this essential text. The book begins with a geometrical study of points, lines, and planes and slowly progresses the reader towards a mastery of screw theory with some cutting-edge results, all while using only basic linear algebra and ordinary vectors. It features a discussion of the geometry of parallel and serial robot manipulators, in addition to the reciprocity of screws and a singularity study. Familiarizing the reader with screw geometry in order to study the statics and kinematics of robots and mechanisms, this is a perfect resource for engineers and graduate students.

Carl D. Crane III is a professor in the department of mechanical and aerospace engineering at the University of Florida. He is a fellow of the ASME.

Michael Griffis is a senior lecturer at the University of Florida.

Joseph Duffy was a great kinematician who passionately promoted screw theory. He was the director of the Center for Intelligent Machines and Robotics at the University of Florida and a graduate research professor, who taught screw theory. He received countless awards, including the ASME Machine Design Award in 2000.

Screw Theory and Its Application to Spatial Robot Manipulators

CARL D. CRANE III
University of Florida

MICHAEL GRIFFIS
University of Florida

JOSEPH DUFFY
University of Florida

CAMBRIDGE
UNIVERSITY PRESS

Shaftesbury Road, Cambridge CB2 8EA, United Kingdom

One Liberty Plaza, 20th Floor, New York, NY 10006, USA

477 Williamstown Road, Port Melbourne, VIC 3207, Australia

314–321, 3rd Floor, Plot 3, Splendor Forum, Jasola District Centre, New Delhi – 110025, India

103 Penang Road, #05–06/07, Visioncrest Commercial, Singapore 238467

Cambridge University Press is part of Cambridge University Press & Assessment, a department of the University of Cambridge.

We share the University's mission to contribute to society through the pursuit of education, learning and research at the highest international levels of excellence.

www.cambridge.org
Information on this title: www.cambridge.org/9780521630894

DOI: 10.1017/9781139019217

First published 2022

A catalogue record for this publication is available from the British Library

ISBN 978-0-521-63089-4   Hardback

To Anne, Sherry, and Karen.

The women who loved us, and who endured us…
while we played with robots.

Joseph, Carl, and Michael

Contents

Preface

Screw theory soared at the University of Florida (UF) when Dr. Duffy arrived in the summer of 1980. The writing of this book started then as class notes for our "Robots 2" course. An original tattered copy of Ball's book (Ball, 1900) was found at the UF library. Historically relevant works such as those by Klein (1939), Plücker (1865), Grassmann (1862), Dimentberg (1968), and Brand (1947) formed our foundation. Early on, Dr. Duffy began collaborating with visitors such as Koichi Sugimoto, and the first papers were published, (Sugimoto and Duffy 1982). Early students such as Harvey Lipkin, Gib Lovell, and Maher Mohamed set the tone and established a hunger for screw theory at UF. It had begun.

As he became the director for the Center for Intelligent Machines and Robotics (CIMAR), Dr. Duffy made it a distinct priority to engage with other screw-theorists around the world. We visited many countries and universities and attended various Romansy, IFToMM, and ASME Mechanisms conferences. Wherever we went, we found the local secondhand math bookstores to buy up whatever historical books we could find.

There was a long period of enlightenment when we hosted renowned kinematicians and geometers in Gainesville. There was such a free and unselfish flow of ideas, a shared passion, and a hearty collaboration. Most enjoyable were the spirited discussions during social hour. The positive influence of these interactions on this book cannot be overstated. For example, Kenneth Hunt's lecture series on screw systems inspired many aspects of this work. In addition to Sugimoto and Hunt, other screw-theory-inspiring visitors included Jack Phillips, John Rees-Jones, Trevor Davies, Eric Primrose, Chris Gibson, Adolf Karger, Vincenzo Parenti Castelli, and David Kerr. We thank those folks, and we thank Erskine Crossley, Bob Bicker, E. A. Dijksman, Lou Torfason, Manfred Hiller, Christoph Woernle, Andres Kecskemethy, and Duncan Marsh for their technical and personal interactions. We also thank all of the others who came, listened, and shared.

Over the 40 years of the course, the CIMAR students have been the lifeblood of this work. This book greatly benefited from students like Harvey, Gib, and Maher. Early days also saw Steve Derby, Bob Freeman, Mark Thomas, Danny Cox, and Keith Soldner. Over the years, we benefited from the contributions of Jose Rico, David Dooner, Lotfi Romdhane, Yong Soeb Chung, Wei Lin, Shannon Ridgeway, Carol Chesney, Phil Adsit, and Byron Knight. We thank those folks, along with Resit Soylu, Bo Zhang, Jeff Wit, Waheed Abbasi, Chang-Hsin Chen, Shih-Chien Chiang, Yong

Authors (L-to-R): Carl, Mike, Joe (1987 at UF)

Je Choi, Jaehoon Lee, Ming-Jer Lin, David Rocheleau, Mark Swinson, Ivan Baiges, Li Young, and Al Nease. We also thank all of the many, many other students who came, and invested in and contributed to parts of this book as well. We believe several thousand students have taken this course over the years at UF (and Dr. Crane has taught this course admirably for many years now – an observation by Mike).

We also thank the UF faculty who endured the many dissertations and gave their insights to help us formulate this volume. These include Gary Matthew, Ralph Selfridge, George Sandor, Keith Doty, and Neil White.

In the following book, Chapter 1 on points, planes, and lines has actually seen very little change since its early days. It forms the introduction for the new student to homogeneous coordinates and their power. Chapter 2 gives the student a solid foundation for coordinate system transformations and how they work on points, lines, planes, and screws. Chapter 3 introduces the cylindroid using the wrench and statics. Chapter 4 applies all of this to the twist and kinematics. It was probably in the 1990s that Dr. Duffy swapped Chapters 3 and 4, as we used to use the twist and kinematics to introduce the cylindroid.

Chapter 5 focuses on reciprocity for the first time and shows the student how to use it to analyze a robot. Chapter 6 empowers the student with the ability to "see" where a singularity can occur by looking at the robot and then analyzing it geometrically. Chapter 7 studies acceleration as found in kinematic serial chains.

The following is presented as a fresh view of "classical screw theory" since it heavily favors geometry but only requires basic algebraic geometry, starting with ordinary vectors and some linear algebra (see Hunt (1978) or Davidson and Hunt (2004) as other examples). For more advanced, group-theory-based presentations, one could consult Selig (2005) or Rico, Gallardo, and Ravani (2003), for example.

In essence, the following gives an engineer's view of how to use screw theory to analyze the kinematics and statics of a robotic manipulator or other kind of mechanism. To the extent we can, we favor tangible, physical, mechanical engineering-type descriptions over precise mathematical proofs or presentations. We strive to put images in the reader's mind of what is physically happening or could happen. As the reader would surely know, there is so much to say on the subject and so many ways to say it. Toward that end, we welcome all comments, and we genuinely hope you enjoy what's to come.

1 Geometry of Points, Lines, and Planes

...and then geometry will become what
geometry ought to be.

Mr. Querulous
Ball's "*A Dynamical Parable*" (1887)

1.1 Introduction

Points, lines, and planes are the fundamental elements of spatial geometry. A point can be thought of as a location in 3D space, and its coordinates have units of length. A line can be considered to be an infinite collection of points defined by a direction (which is a dimensionless vector) that passes through some given point (which has units of meters). A plane is a two-dimensional set of points that can be defined, for example, by three points or by a line and one point. This chapter introduces the concept of homogeneous coordinates as applied to points, lines, and planes. The *homogeneous coordinates* of each will be defined together with the *equation* for each. The equation of a point, line, or plane will be shown to be a vector equation where any vector that satisfies that equation is a member of that point, line, or plane.

1.2 The Position Vector of a Point

The position vector to a point Q_1 from a reference point O will be referred to as r_1 and can be expressed in the form

$$r_1 = \frac{x_1\, i + y_1\, j + z_1\, k}{w_1} \tag{1.1}$$

or

$$r_1 w_1 = S_{O1}, \tag{1.2}$$

where $S_{O1} = x_1\, i + y_1\, j + z_1\, k$ and the components of the vector S_{O1} have units of length. The term w_1 is dimensionless. In Figure 1.1 it is assumed that $w_1 = 1$ and (x_1, y_1, z_1) are the usual Cartesian coordinates for the point Q_1. The coordinates r of some general point Q may be expressed as

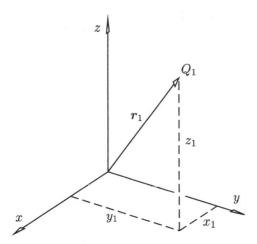

Figure 1.1 Coordinates of a point

$$r w = S_O, \qquad (1.3)$$

where $S_O = x\,\boldsymbol{i} + y\,\boldsymbol{j} + z\,\boldsymbol{k}$. The subscript O has been introduced to signify that S_O is origin dependent. Clearly, if we choose some other reference point, the actual point Q would not change. However, the coordinates (x, y, z), which determine Q, would change. The ratios x/w, y/w, and z/w are three independent scalars and, therefore, there are ∞^3 points in space.

It is interesting to consider the cases where $S_O = \mathbf{0}$ and where $w = 0$. From (1.3)

$$|r| = \frac{|S_O|}{|w|}, \qquad (1.4)$$

where the notation $||$ denotes absolute magnitude. From (1.4), when $|S_O| = 0$, $|r| = 0$ and Q coincides with the reference point O. When $|w| = 0$, $|r|$ is infinite and the point Q is said to be at infinity in the direction parallel to S_O. The introduction of w makes it possible to designate a point by the array of four coordinates $(w; x, y, z)$[1] and is a means of introducing the concept of infinity, or more specifically infinite points, into the geometry without introducing the symbol ∞. Any point at infinity is designated by the coordinates $(0; x, y, z)$. It is important to recognize that $|S_O|$ and $|w|$ cannot be zero simultaneously since $|r|$ would be indeterminate. In other words, the array $(0; 0, 0, 0)$ is not permitted.

It is of interest to examine the geometry of points labeled with the four coordinates $(w; x, y, z)$. A number of readers will know that this geometry is called projective geometry, which is the subject of many texts (see Coxeter, Meserve, Semple and Knee-bone, Faulkner, and Scott to name a few.[2]) Firstly, the four coordinates $(w; x, y, z)$ are homogeneous since from (1.3) $(\lambda w; \lambda S_O)$, where λ is a non-zero scalar, determine the

---

[1] The semi-colon is introduced into the notation to signify that the dimension of w is different from that of x, y, and z.

[2] Coxeter (2003), Meserve (2010), Semple and Kneebone (1998), Faulkner (2006), and Scott (1894).

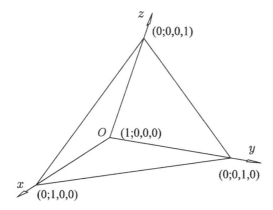

Figure 1.2 Tetrahedron of reference

same point. For instance, the coordinates $(1; 2, 3, 4)$, $(2; 4, 6, 8)$, and $(-2; -4, -6, -8)$, where the last three coordinates have the same unit of length, all determine the same point. Further, the homogeneous coordinates for the origin of this coordinate system are $(w; 0, 0, 0)$ or $(1; 0, 0, 0)$. The homogeneous coordinates for points on the x, y, and z axes are, respectively, $(w; x, 0, 0)$, $(w; 0, y, 0)$, and $(w; 0, 0, z)$. Therefore, the homogeneous coordinates for points at infinity on the x, y, and z axes are $(0; x, 0, 0)$, $(0; 0, y, 0)$, and $(0; 0, 0, z)$ or $(0; 1, 0, 0)$, $(0; 0, 1, 0)$, and $(0; 0, 0, 1)$ or $(0; -1, 0, 0)$, $(0; 0, -1, 0)$, and $(0; 0, 0, -1)$. These three points together with the origin form the four vertices of the so-called tetrahedron of reference, illustrated in Figure 1.2.

The projective space of Figure 1.2 is radically different from the Euclidean space labeled by the x, y, and z Cartesian coordinate frame in Figure 1.1. There can be no calibration of the x, y, and z axes in Figure 1.2, i.e., there is no concept of a unit length of measure or of angle. The four coordinates $(w; x, y, z)$ have no dimensions, and the semi-colon is somewhat redundant in projective space. It will, however, be retained simply to signify the order in which the coordinates are written. There are no parallel lines that do not meet. Lines that are parallel in the Euclidean sense meet in the projective space at points on the plane at infinity, which can be drawn through the three points $(0; 1, 0, 0)$, $(0; 0, 1, 0)$, and $(0; 0, 0, 1)$ at infinity on the x, y, and z axes. This brief discussion is sufficient for the purposes of this text. It remains to label the four planes of the tetrahedron of reference with their homogeneous plane coordinates and to label the six edges of the tetrahedron with their homogeneous line coordinates. The homogeneous coordinates of planes and lines are defined in the subsequent two sections of this chapter.

1.3   The Equation of a Plane

The equation of a plane through a point Q_1 with coordinates (x_1, y_1, z_1) and perpendicular to a vector $S = A\,\boldsymbol{i} + B\,\boldsymbol{j} + C\,\boldsymbol{k}$ (see Figure 1.3) can be expressed in the form

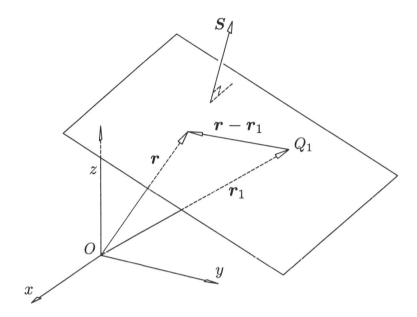

Figure 1.3 Determination of a plane

$$(r - r_1) \cdot S = 0, \tag{1.5}$$

where $r = x\,i + y\,j + z\,k$ is a vector from the origin to any general point on the plane, and r_1 is the vector from the origin to the point Q_1. The components of the vectors r and r_1 have units of length, while the components of S are dimensionless. Equation (1.5) may be written as

$$r \cdot S + D_O = A\,x + B\,y + C\,z + D_O = 0, \tag{1.6}$$

where

$$D_O = -r_1 \cdot S = -(A\,x_1 + B\,y_1 + C\,z_1). \tag{1.7}$$

The scalar value D_O has units of length and is origin dependent.

The coordinate of a plane will be written as $[D_O; A, B, C]$, where square brackets are now used to distinguish these values from the coordinates of a point where parentheses were used. The array

$$[D_O; A, B, C]$$

represents the homogeneous coordinates of the plane since from (1.6) the coordinates of

$$[\lambda D_O; \lambda A, \lambda B, \lambda C],$$

where λ is any non-zero scalar, determine the same plane. From (1.6) it is apparent that the dimension of D_O is different from those of A, B, and C, and a semi-colon is introduced in the array to signify this.

Dividing (1.6) by D_O yields

$$\frac{A}{D_O}x + \frac{B}{D_O}y + \frac{C}{D_O}z + 1 = 0. \tag{1.8}$$

The ratios $\frac{A}{D_O}$, $\frac{B}{D_O}$, and $\frac{C}{D_O}$ are three independent scalars and, therefore, there are ∞^3 planes in space.

The distance of the plane from the reference point O is determined by the length of the vector $\boldsymbol{p}$, which extends from point O to a point on the plane such that $\boldsymbol{p}$ is perpendicular to the plane. Since the vectors $\boldsymbol{p}$ and $\boldsymbol{S}$ must be parallel,

$$\boldsymbol{p} \times \boldsymbol{S} = 0. \tag{1.9}$$

Further, since $\boldsymbol{p}$ is a vector to a point on the plane, it must satisfy (1.6) and

$$\boldsymbol{p} \cdot \boldsymbol{S} = -D_O. \tag{1.10}$$

Performing a cross product of $\boldsymbol{S}$ with (1.9) yields

$$\boldsymbol{S} \times (\boldsymbol{p} \times \boldsymbol{S}) = \boldsymbol{0}. \tag{1.11}$$

Expanding this expression[3] yields

$$\boldsymbol{p}\,(\boldsymbol{S} \cdot \boldsymbol{S}) - \boldsymbol{S}\,(\boldsymbol{p} \cdot \boldsymbol{S}) = \boldsymbol{0}. \tag{1.12}$$

Using (1.10) to substitute for $\boldsymbol{p} \cdot \boldsymbol{S}$ and then solving for $\boldsymbol{p}$ gives

$$\boldsymbol{p} = \frac{-D_O \boldsymbol{S}}{\boldsymbol{S} \cdot \boldsymbol{S}}. \tag{1.13}$$

The perpendicular distance of the plane from the origin, i.e., the magnitude of $\boldsymbol{p}$ is, therefore,

$$|\boldsymbol{p}| = \frac{|-D_O|\ |\boldsymbol{S}|}{|\boldsymbol{S}|\ |\boldsymbol{S}|} = \frac{|-D_O|}{|\boldsymbol{S}|}. \tag{1.14}$$

Clearly, when $|\boldsymbol{S}| = 1$, the triple (A, B, C) are the direction cosines of a vector normal to the plane, and $|D_O|$ is the perpendicular distance of the plane from the reference point O. Further, when $D_O = 0$, the equation of the plane, i.e., (1.6), becomes

$$A x + B y + C z = 0, \tag{1.15}$$

and the plane passes through O.

The yz plane passes through the origin, and its normal vector is parallel to the x axis. The coordinates for this plane are, thus, $[0; A, 0, 0]$ or $[0; 1, 0, 0]$. Similarly, the coordinates for the zx plane and the xy plane are $[0; 0, 1, 0]$ and $[0; 0, 0, 1]$, respectively. These three planes together with the plane at infinity for which $\boldsymbol{S} = \boldsymbol{0}$ and whose coordinates[4] are therefore $[D; 0, 0, 0]$ or $[1; 0, 0, 0]$ are labeled in Figure 1.4. All points at infinity lie on this plane.

---

[3] The product $\boldsymbol{a} \times (\boldsymbol{b} \times \boldsymbol{c})$ is equal to $(\boldsymbol{a} \cdot \boldsymbol{c})\boldsymbol{b} - (\boldsymbol{a} \cdot \boldsymbol{b})\boldsymbol{c}$.
[4] From here on the subscript O will be omitted from, D, although it should be remembered that the value of D is, indeed, origin dependent.

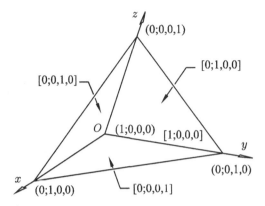

Figure 1.4 Planes and points on the tetrahedron of reference

The plane at infinity, $[1; 0, 0, 0]$, can also be thought of as the outer surface of a sphere of infinite radius centered at the origin of the reference system. This surface, with its infinite radius of curvature, would appear as a plane to an observer located at a point on this plane.

It is important to recognize that in three-dimensional projective space a point and a plane are analogous, or, more specifically, they are *dual*. Sets of four homogeneous coordinates $(w; x, y, z)$ and $[D; A, B, C]$ define points and planes in projective space, and an ∞^3 of both points and planes fill space. A plane can be drawn through three non-collinear points. This statement can be rephrased for a point by making the appropriate grammatical changes in order for it to make sense. The dual statement is that three non-parallel planes meet or intersect at a point. It is always possible to formulate (prove) a proposition (theorem) for one dual element and to simply state a corresponding proposition (theorem) for the corresponding dual element. A further two examples are *a line is the join of two points* which is dual to *a line is the meet of two planes* and *a line intersects a plane (which does not contain the line) in a point* which is dual to *a line and any point not on the line determine a plane*.

Finally, Klein considered the point and the plane to be equally important (see Klein [1939]). One can write their incidence relationship in the form

$$D\,w + A\,x + B\,y + C\,z = 0. \tag{1.16}$$

The point coordinates $(w; x, y, z)$ and plane coordinates $[D; A, B, C]$ play equal roles in (1.16). When the coordinates $[D; A, B, C]$ are specified, (1.16) expresses the condition that ∞^2 points lie on a plane. When the coordinates $(w; x, y, z)$ are specified, (1.16) expresses the condition that an ∞^2 (a bundle) of planes passes through the point.

1.3.1   Sample Problem

The coordinates of a plane are given as $[-8\,;2,\,-3,\,5]$, where the first term has units of meters and the last three are dimensionless. Determine the coordinates of a point on the plane.

The point to be determined will be written as

$$\boldsymbol{r} = r_x\,\mathbf{i} + r_y\,\mathbf{j} + r_z\,\mathbf{k}. \tag{1.17}$$

This point must satisfy the equation of the plane and, thus,

$$\boldsymbol{r}\cdot\boldsymbol{S} + D_O = 0, \tag{1.18}$$

$$2\,r_x - 3\,r_y + 5\,r_z - 8 = 0. \tag{1.19}$$

Equation (1.19) is one equation in three unknowns. Free choices may be made for two of the point coordinate values, and the third may then determined from (1.19). For example, choosing $r_x = 0$ and $r_y = -1$ and solving for r_z gives $r_z = 1$ and, thus, the point $[1; 0, -1, 1]$, where the first term is dimensionless and the last three have units of meters, is on the plane.

1.3.2   Sample Problem

(i) *Determine the coordinates of the plane that passes through the three points* $P = (1; 3, 4, 1)$, $Q = (1; -1, 2, 4)$, *and* $R = (1; 3, 2, 2)$, *where the Cartesian coordinates are give in units of meters.*

The direction vector from point P to point Q may be written as

$$\boldsymbol{S}_{pq} = -4\,\boldsymbol{i} - 2\,\boldsymbol{j} + 3\,\boldsymbol{k}, \tag{1.20}$$

and the direction vector form point P to point R may be written as

$$\boldsymbol{S}_{pr} = -2\,\boldsymbol{j} + 1\,\boldsymbol{k}, \tag{1.21}$$

where these direction vectors are dimensionless. The vector perpendicular to the plane, $\boldsymbol{S}$, may be calculated as

$$\boldsymbol{S} = \boldsymbol{S}_{pq} \times \boldsymbol{S}_{pr} = 4\,\boldsymbol{i} + 4\,\boldsymbol{j} + 8\,\boldsymbol{k}, \tag{1.22}$$

where again the direction vector is dimensionless. From (1.7)

$$D_O = -\boldsymbol{r}_1 \cdot \boldsymbol{S}, \tag{1.23}$$

where $\boldsymbol{r}_1$ can be any of the three points on the plane. Using point P gives

$$D_O = -(3\boldsymbol{i} + 4\boldsymbol{j} + 1\boldsymbol{k}) \cdot (4\boldsymbol{i} + 4\boldsymbol{j} + 8\boldsymbol{k}) = -36 \text{ m}. \tag{1.24}$$

Thus, the coordinates of the plane may be written as

$$[D_O; \boldsymbol{S}] = [-36; 4, 4, 8], \tag{1.25}$$

where the first component has units of meters and the remaining three are dimensionless.

(ii) *Determine the distance of this plane from the origin.*

The vector from the origin that is perpendicular to the plane can be determined from (1.13) as

$$p = \frac{-D_O S}{S \cdot S} = \frac{36 (4i + 4j + 8k)}{4^2 + 4^2 + 8^2} = 1.5i + 1.5j + 3k, \qquad (1.26)$$

where the components of the vector have units of meters. The distance of the plane from the origin is equal to the magnitude of this vector, i.e.,

$$|p| = 3.674 \text{ m}. \qquad (1.27)$$

(iii) *Determine the Z coordinate of the point on the plane whose x and y coordinate values are 6 m and −10 m, respectively.*

The point on the plane can be written as

$$r_4 = 6i - 10j + z_4 k. \qquad (1.28)$$

Inserting this point into (1.6), the equation of the plane, yields

$$r_4 \cdot S + D_O = 0,$$

$$(6i - 10j + z_4 k) \cdot (4i + 4j + 8k) - 36 = 0. \qquad (1.29)$$

Solving for z_4 gives

$$z_4 = 6.5 \text{ m}. \qquad (1.30)$$

1.4   Projection of a Point onto a Plane

Often it is desired to find the point on a plane that is closest to a given point. Figure 1.5 shows a plane whose coordinates are given as $[D_O; S]$ and a given point Q_1 whose coordinates are defined by the vector r_1. The objective is to determine the coordinates of the point Q_p, which is the point on the plane that is closest to the point Q_1. The coordinates of point Q_p are defined by the vector r_p.

The vector from point Q_p to Q_1 is labeled as $r_{p \to 1}$, and it is apparent that this vector must be parallel to S. Thus, $r_{p \to 1}$ is some scalar multiple of S and can be written as

$$r_{p \to 1} = d\, S. \qquad (1.31)$$

Further it can be seen from the figure that

$$r_p = r_1 - r_{p \to 1} = r_1 - d\, S. \qquad (1.32)$$

Since point Q_p lies in the given plane, the vector r_p must satisfy the equation of the plane and, thus,

$$(r_1 - d\, S) \cdot S + D_O = 0. \qquad (1.33)$$

Solving for d gives

$$d = \frac{r_1 \cdot S + D_O}{S \cdot S}. \qquad (1.34)$$

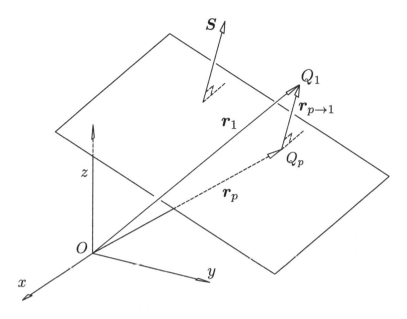

Figure 1.5 Projection of a point onto a plane

Substituting (1.34) into (1.32) gives the result

$$r_p = r_1 - \left(\frac{r_1 \cdot S + D_O}{S \cdot S}\right) S. \tag{1.35}$$

It is of note that the distance from point Q_1 to the plane is obtained as the magnitude of (1.31) as

$$|r_{p \to 1}| = d\,|S|, \tag{1.36}$$

and a positive value for d indicates that the point lies on the side of the plane pointed to by the direction of S.

1.5 The Equation of a Line

The join of two distinct points r_1 (x_1, y_1, z_1) and r_2 (x_2, y_2, z_2), where the elements of r_1 and r_2 have units of length, determine a line. The vector S whose direction is along the line may be written as

$$S = r_2 - r_1. \tag{1.37}$$

Direction is a unitless concept and, thus, the elements of S are dimensionless. The vector S may alternatively be expressed as

$$S = L\,i + M\,j + N\,k, \tag{1.38}$$

where $L = x_2 - x_1$, $M = y_2 - y_1$, and $N = z_2 - z_1$ are defined as the dimensionless direction ratios. From (1.38) the direction ratios (L, M, N) are related to $|S|$ by

$$L^2 + M^2 + N^2 = |S|^2. \tag{1.39}$$

Often, L, M, and N are expressed in the form

$$L = \frac{x_2 - x_1}{|S|}, \qquad M = \frac{y_2 - y_1}{|S|}, \qquad N = \frac{z_2 - z_1}{|S|}, \tag{1.40}$$

which are unit direction ratios or direction cosines of the line. In this case, (1.39) reduces to

$$L^2 + M^2 + N^2 = 1. \tag{1.41}$$

Letting r designate a vector from the origin to any general point on the line (see Figure 1.6), it is apparent that the vector $r - r_1$ is parallel to S. Thus, it may be written that

$$(r - r_1) \times S = 0. \tag{1.42}$$

This can be expressed in the form

$$r \times S = S_{OL}, \tag{1.43}$$

where

$$S_{OL} = r_1 \times S \tag{1.44}$$

is the moment of the line about the origin O, which is clearly origin dependent. The elements of the vector S_{OL} have units of length. Further, since $S_{OL} = r_1 \times S$, the vectors S and S_{OL} are perpendicular and, as such, satisfy the orthogonality condition

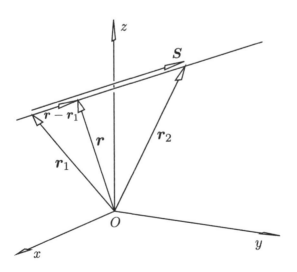

Figure 1.6 Determination of a line

$$S \cdot S_{OL} = 0. \tag{1.45}$$

The coordinates of a line will be written as $\{S; S_{OL}\}$ and will be referred to as the Plücker coordinates[5] of the line. The semi-colon is introduced to signify that the dimensions differ between the first three coordinates and the last three. The coordinates $\{S; S_{OL}\}$ are homogeneous since from (1.43) the coordinates $\{\lambda S; \lambda S_{OL}\}$, where λ is a non-zero scalar, determine the same line.

Expanding (1.44) yields

$$S_{OL} = \begin{vmatrix} i & j & k \\ x_1 & y_1 & z_1 \\ L & M & N \end{vmatrix}, \tag{1.46}$$

which can be expressed in the form

$$S_{OL} = P\,i + Q\,j + R\,k, \tag{1.47}$$

where

$$P = y_1\,N - z_1\,M, \tag{1.48}$$
$$Q = z_1\,L - x_1\,N,$$
$$R = x_1\,M - y_1\,L.$$

From (1.38) and (1.47) the orthogonality condition $S \cdot S_{OL} = 0$ can be expressed in the form

$$L\,P + M\,Q + N\,R = 0. \tag{1.49}$$

The six Plücker coordinates of the line $\{L, M, N; P, Q, R\}$ are illustrated in Figure 1.7. Note that for the case shown in the figure, L will have a negative value. Unitized coordinates for a line can be obtained by imposing the constraint that $|S| = 1$. The Plücker coordinates must thus satisfy equations (1.41) and (1.49) and, therefore, only four of the six scalars L, M, N, P, Q, and R are independent. It follows that there are ∞^4 lines in space.[6]

Equations (1.37) and (1.44) can be used to obtain the Plücker coordinates of a line when given two points on the line. It is also important to be able to determine a point on the line when given the Plücker coordinates of the line. Suppose that the Plücker coordinates $\{L, M, N; P, Q, R\}$ of a line are given. Let $r = x\,i + y\,j + z\,k$ represent a vector to some point on the line. Thus, r must satisfy (1.43) and

$$(x\,i + y\,j + z\,k) \times (L\,i + M\,j + N\,k) = P\,i + Q\,j + R\,k. \tag{1.50}$$

Restating the cross product on the left side of this equation gives

$$\begin{vmatrix} i & j & k \\ x & y & z \\ L & M & N \end{vmatrix} = P\,i + Q\,j + R\,k. \tag{1.51}$$

---

[5] See Plücker (1865).

[6] Systems of lines and their properties are described in Hunt (1978) (pp. 310–330), which contains an extensive bibliography on the subject. A line series (∞^1), congruence (∞^2), and complex (∞^3) are discussed.

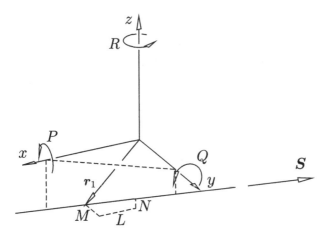

Figure 1.7 Plücker line coordinates

Equating the components of the i, j, and k vectors yields the three scalar equations

$$y N - z M = P, \tag{1.52}$$
$$z L - x N = Q,$$
$$x M - y L = R,$$

which may be written in matrix format as

$$\begin{bmatrix} 0 & N & -M \\ -N & 0 & L \\ M & -L & 0 \end{bmatrix} \begin{bmatrix} x \\ y \\ z \end{bmatrix} = \begin{bmatrix} P \\ Q \\ R \end{bmatrix}, \tag{1.53}$$

where the 3 × 3 coefficient matrix is skew symmetric. It can be shown that the rank of this coefficient matrix is two and the equations of (1.52) are, therefore, linearly dependent. As such, an infinite number of solutions exist for x, y, and z (corresponding to the infinite number of points on the line $\{L, M, N; P, Q, R\}$). Any arbitrary value for x may be selected, and corresponding values for y and z may be determined from the last two equations of (1.52) as

$$y = \frac{x M - R}{L}, \quad z = \frac{x N + Q}{L}. \tag{1.54}$$

Note that if $L = 0$, then x is a constant given by either the second or third equation of (1.52). The value for y or z could then be arbitrarily selected and the other calculated from the first equation of (1.52).

The distance of a line from the origin is determined by the length of the vector p, which originates at O and terminates at a point on the line such that the direction of p is perpendicular to the direction of the line, S. The vector p must also satisfy (1.43) and, therefore,

$$p \cdot S = 0, \tag{1.55}$$

$$p \times S = S_{OL}. \tag{1.56}$$

Performing a cross product of S with (1.56) yields

$$S \times (p \times S) = S \times S_{OL}. \tag{1.57}$$

Expanding the left side of (1.57) gives

$$(S \cdot S)p - (p \cdot S)S = S \times S_{OL}. \tag{1.58}$$

Since p is perpendicular to S, $p \cdot S = 0$. Substituting this into (1.58) and solving for p gives

$$p = \frac{S \times S_{OL}}{S \cdot S}. \tag{1.59}$$

A vector e is now defined as a unit vector perpendicular to S and S_{OL} and may be written as

$$e = \frac{S \times S_{OL}}{|S \times S_{OL}|}. \tag{1.60}$$

Substituting $(S \times S_{OL}) = |S \times S_{OL}|e$ in (1.59) yields

$$p = \frac{|S \times S_{OL}|}{S \cdot S}e. \tag{1.61}$$

The magnitude of the cross product in the numerator of (1.61) is simply $|S||S_{OL}| \sin \frac{\pi}{2}$. The scalar product in the denominator will equal the square of the magnitude of S. Equation (1.61) may, thus, be written as

$$p = \frac{|S||S_{OL}|}{|S||S|}e = \frac{|S_{OL}|}{|S|}e. \tag{1.62}$$

Therefore, the distance of the line from the origin may be determined as

$$|p| = \frac{|S_{OL}|}{|S|}. \tag{1.63}$$

When $S_{OL} = 0$, $|p| = 0$ and the line passes through the origin and its coordinates are $\{S; 0\}$. When $S = 0$, $|p| = \infty$ and the line is a line at infinity. In this case, the direction of the moment vector defines the line at infinity, and its coordinates are written as $\{0; S\}$, where S is dimensionless. Further, the coordinates of the line at infinity are origin independent. This line at infinity lies in the plane at infinity. If the plane at infinity is thought of as the surface of a sphere of infinite radius, then a line in the plane at infinity can be thought of as a circle on this sphere. For the line $\{0; S\}$ which lies at infinity, the direction of the moment, S, must be perpendicular to the direction of the line, as is the case for all lines (see Figure 1.8). Thus, for the line $\{0; S\}$ the direction of the moment vector S can be thought of as being perpendicular to the "plane" defined by the "circle of infinite radius".

The Plücker coordinates for the line joining the points with coordinates $(1; x_1, y_1, z_1)$ and $(1; x_2, y_2, z_2)$ were elegantly expressed by Grassmann[7] by the six 2×2 determinants of the array

---

[7] On pg. 20, Klein (1939) gave immense credit to Grassmann (1862) regarding his extension theory.

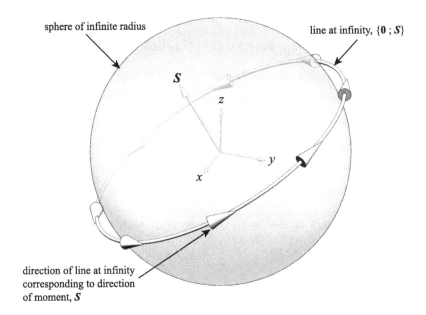

sphere of infinite radius

line at infinity, $\{0 ; S\}$

S

z

y

x

direction of line at infinity
corresponding to direction
of moment, S

Figure 1.8 Conceptualization of the line at infinity, $\{0; S\}$.

$$\begin{bmatrix} 1 & x_1 & y_1 & z_1 \\ 1 & x_2 & y_2 & z_2 \end{bmatrix} \tag{1.64}$$

as

$$L = \begin{vmatrix} 1 & x_1 \\ 1 & x_2 \end{vmatrix}, \quad M = \begin{vmatrix} 1 & y_1 \\ 1 & y_2 \end{vmatrix}, \quad N = \begin{vmatrix} 1 & z_1 \\ 1 & z_2 \end{vmatrix},$$

$$P = \begin{vmatrix} y_1 & z_1 \\ y_2 & z_2 \end{vmatrix}, \quad Q = \begin{vmatrix} z_1 & x_1 \\ z_2 & x_2 \end{vmatrix}, \quad R = \begin{vmatrix} x_1 & y_1 \\ x_2 & y_2 \end{vmatrix}. \tag{1.65}$$

It is a useful exercise to deduce the Plücker coordinates of the six edges of the tetrahedron of reference using (1.65), using the pairs of coordinates labeling the vertices
of the tetrahedron of reference shown in Figure 1.2.

1.5.1   Sample Problem

(i) *Determine the Plücker coordinates of the line that passes through the points*
$r_1 = 3i + 5j - 6k$ *and* $r_2 = 6i - 5j + 2k$, *where the point coordinates are given in*
units of meters.

The direction of the line is obtained from (1.37) as

$$S = r_2 - r_1 = 3i - 10j + 8k, \tag{1.66}$$

where the elements of the direction vector S are dimensionless. The moment of the line is calculated from (1.44) as

$$S_{OL} = r_1 \times S = -20i - 42j - 45k, \tag{1.67}$$

where the elements of S_{OL} have units of meters. The Plücker coordinates of the line may now be written as

$$\{3, -10, 8; -20 \text{ m}, -42 \text{ m}, -45 \text{ m}\}. \tag{1.68}$$

(ii) *Determine the perpendicular distance of this line from the origin.*

The coordinates of the point on the line that is closest to the origin can be determined from (1.59) as

$$p = \frac{S \times S_{OL}}{S \cdot S} = \frac{(3i - 10j + 8k) \times (-20i - 42j - 45k)}{(3i - 10j + 8k) \cdot (3i - 10j + 8k)} \tag{1.69}$$
$$= 4.543i - 0.145j - 1.884k,$$

where the elements of the vector p have units of meters. The magnitude of p is the distance of the perpendicular distance of the line from the origin and is calculated as

$$|p| = 4.921 \text{ m}. \tag{1.70}$$

(iii) *Determine the coordinates of another point on the line.*

An arbitrary point on the line will be referred to as r, where

$$r = xi + yj + zk. \tag{1.71}$$

The equation of the line was presented in (1.43) and may be written as

$$r \times S = S_{OL}, \tag{1.72}$$
$$(xi + yj + zk) \times (3i - 10j + 8k) = -20i - 42j + 45k,$$

where the terms on the right side of the equation and the unknowns x, y, and z have units of meters. Expanding the cross product on the left hand side of this equation gives

$$(8y + 10z)i + (3z - 8x)j + (-10x - 3y)k = -20i - 42j + 45k. \tag{1.73}$$

Equating the i, j, and k components of this equation yields the three scalar equations

$$8y + 10z = -20, \tag{1.74}$$
$$3z - 8x = -42,$$
$$-10x - 3y = -45.$$

Multiplying the first equation by 3 and the second equation by -10 and adding gives

$$24y + 80x = 360, \tag{1.75}$$

which, when divided by -8, is identical to the third equation of (1.74) and, thus, the three scalar equations are linearly dependent. This was to be expected, since there are an infinity of points on a line. A free choice can be made for one of the coordinates, say $x = 0$, and the other two parameters calculated. In this case, $y = 15$ m and $z = -14$ m.

1.6   Two Planes Determine a Line

The coordinates of two planes are given as $[D_{O1}; A_1, B_1, C_1]$ and $[D_{O2}; A_2, B_2, C_2]$. From (1.6), the equations of each plane may be written as

$$r_1 \cdot S_1 + D_{O1} = 0, \tag{1.76}$$

$$r_2 \cdot S_2 + D_{O2} = 0. \tag{1.77}$$

where r_1 and r_2 are vectors to any point on the first and second plane, respectively, and

$$S_1 = A_1 i + B_1 j + C_1 k, \tag{1.78}$$

$$S_2 = A_2 i + B_2 j + C_2 k. \tag{1.79}$$

The equation of the line of intersection of these two planes (see Figure 1.9) will now be determined.

The line of intersection is perpendicular to each of the vectors S_1 and S_2 and it is, therefore, parallel to $S_1 \times S_2$. Expanding the triple vector cross product $r \times (S_1 \times S_2)$, where r is a vector to any point on the line of intersection, yields the vector equation for the line and is written as

$$r \times (S_1 \times S_2) = (r \cdot S_2)S_1 - (r \cdot S_1)S_2. \tag{1.80}$$

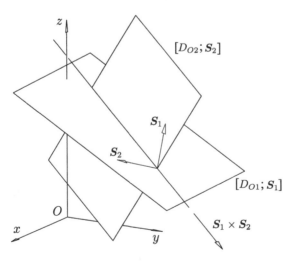

Figure 1.9 Line of intersection of two planes

Since r must lie on both of the planes, (1.76) and (1.77) may be substituted into (1.80) to give

$$r \times (S_1 \times S_2) = (-D_{O2})S_1 - (-D_{O1})S_2. \tag{1.81}$$

The coordinates of the line of intersection of the two planes are obtained from (1.81) as $\{S_1 \times S_2; D_{O1} S_2 - D_{O2} S_1\}$. It is apparent that these coordinates do, indeed, represent a line, as the direction of the line, $S_1 \times S_2$, is perpendicular to the moment of the line, $D_{O1} S_2 - D_{O2} S_1$. When $S_1 \times S_2 = 0$, the planes are parallel, and they intersect in a line in the plane at infinity with coordinates $\{0; D_{O1} S_2 - D_{O2} S_1\}$. Since in this case $S_1 \parallel S_2$, the line at infinity may simply be written as $\{0; S_1\}$.

By substituting (1.78) and (1.79) into the Plücker coordinates of the line of intersection and evaluating the cross product $S_1 \times S_2$, these coordinates may be written as $\{L, M, N; P, Q, R\}$, where

$$L = B_1 C_2 - B_2 C_1, \quad M = C_1 A_2 - C_2 A_1, \quad N = A_1 B_2 - A_2 B_1, \tag{1.82}$$
$$P = D_{O1} A_2 - D_{O2} A_1, \quad Q = D_{O1} B_2 - D_{O2} B_1, \quad R = D_{O1} C_2 - D_{O2} C_1.$$

The Plücker coordinates in (1.82) may be obtained directly using Grassmann's determinant principle by expressing the coordinates of the planes in the 2×4 array

$$\begin{bmatrix} D_{O1} & A_1 & B_1 & C_1 \\ D_{O2} & A_2 & B_2 & C_2 \end{bmatrix} \tag{1.83}$$

and by expanding the sequence of determinants

$$P = \begin{vmatrix} D_{O1} & A_1 \\ D_{O2} & A_2 \end{vmatrix}, \quad Q = \begin{vmatrix} D_{O1} & B_1 \\ D_{O2} & B_2 \end{vmatrix}, \quad R = \begin{vmatrix} D_{O1} & C_1 \\ D_{O2} & C_2 \end{vmatrix}, \tag{1.84}$$
$$L = \begin{vmatrix} B_1 & C_1 \\ B_2 & C_2 \end{vmatrix}, \quad M = \begin{vmatrix} C_1 & A_1 \\ C_2 & A_2 \end{vmatrix}, \quad N = \begin{vmatrix} A_1 & B_1 \\ A_2 & B_2 \end{vmatrix}.$$

The array $\{P, Q, R; L, M, N\}$ is known as the *axis coordinates* for the line determined by the meet of two planes. Clearly, the line can be considered as the *axis* of a pencil of planes. On the other hand, the array $\{L, M, N; P, Q, R\}$ is known as the *ray coordinates* for a line. Clearly, the line can be considered as a *ray* of light joining any two distinct points on the line. The line can thus be formed by pairs of dual elements, i.e., points or planes. Because of this, a line in three-dimensional space is considered to be dual with itself or self-dual.

A simple notation will now be introduced that will be used to distinguish between the ray and axis coordinates in later chapters. The ray and axis coordinates will be designated by lower and upper case symbols $\hat{s} = \{S; S_{OL}\}$ and $\hat{S} = \{S_{OL}; S\}$, respectively. Also, a line, whether written in ray or axis coordinates, will be designated as \$. When the same line \$ is determined by the meet of two planes $[D_{O1}; A_1, B_1, C_1]$ and $[D_{O2}; A_2, B_2, C_2]$ and by the join of two points $(1; x_1, y_1, z_1)$ and $(1; x_2, y_2, z_2)$, as shown in Figure 1.10, then axis coordinates $\{S_{OL}; S\}$ are obtained by counting the 2×2 determinants of

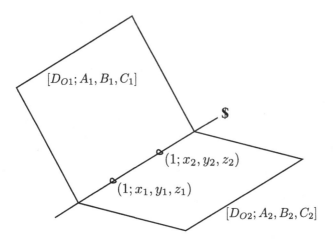

Figure 1.10 A line determined by two planes or two points

$$\begin{bmatrix} D_{O1} & A_1 & B_1 & C_1 \\ D_{O2} & A_2 & B_2 & C_2 \end{bmatrix}, \tag{1.85}$$

whereas the ray coordinates $\{S; S_{OL}\}$ are obtained by counting the 2×2 determinants of

$$\begin{bmatrix} 1 & x_1 & y_1 & z_1 \\ 1 & x_2 & y_2 & z_2 \end{bmatrix}. \tag{1.86}$$

The two sets of coordinates are identical to a scalar multiple σ and

$$\hat{s} = \sigma \mathbf{\Delta} \hat{S}, \tag{1.87}$$

where $\mathbf{\Delta}$ is a 6×6 matrix that converts axis coordinates to ray coordinates and vice versa and is given by

$$\mathbf{\Delta} = \begin{bmatrix} 0 & 0 & 0 & 1 & 0 & 0 \\ 0 & 0 & 0 & 0 & 1 & 0 \\ 0 & 0 & 0 & 0 & 0 & 1 \\ 1 & 0 & 0 & 0 & 0 & 0 \\ 0 & 1 & 0 & 0 & 0 & 0 \\ 0 & 0 & 1 & 0 & 0 & 0 \end{bmatrix}. \tag{1.88}$$

The matrix $\mathbf{\Delta}$ essentially exchanges the first three elements with the last three elements of the column vector $\hat{S}$ and is often written as

$$\mathbf{\Delta} = \begin{bmatrix} \mathbf{0} & I_3 \\ I_3 & \mathbf{0} \end{bmatrix}, \tag{1.89}$$

where, in this case, $\mathbf{0}$ represents a 3×3 matrix with all terms equal to zero, and I_3 represents a 3×3 identity matrix.

As an example, consider a pair of planes with the coordinates $[1; -4, 12, 3]$ and $[1; 3, 4, -12]$, where the first component has units of meters and the last three components are dimensionless. The axis coordinates for the line of intersection $ \$ $ are obtained from the 2×2 determinants of the array

$$\begin{bmatrix} 1 & -4 & 12 & 3 \\ 1 & 3 & 4 & -12 \end{bmatrix} \tag{1.90}$$

and

$$P = \begin{vmatrix} 1 & -4 \\ 1 & 3 \end{vmatrix} = 7 \text{ m}, \quad Q = \begin{vmatrix} 1 & 12 \\ 1 & 4 \end{vmatrix} = -8 \text{ m}, \tag{1.91}$$

$$R = \begin{vmatrix} 1 & 3 \\ 1 & -12 \end{vmatrix} = -15 \text{ m}, \quad L = \begin{vmatrix} 12 & 3 \\ 4 & -12 \end{vmatrix} = -156,$$

$$M = \begin{vmatrix} 3 & -4 \\ -12 & 3 \end{vmatrix} = -39, \quad N = \begin{vmatrix} -4 & 12 \\ 3 & 4 \end{vmatrix} = -52.$$

The numerical results can be verified by using the orthogonality condition of (1.49). The axis coordinates of the line of intersection may thus be written as $\hat{S} = \{S_{OL}; S\}$, where

$$S_{OL} = (7i - 8j - 15k) \text{ m} \tag{1.92}$$

$$S = -156i - 39j - 52k.$$

Two points on this line will now be determined. Writing a point on the line as $r = xi + yj + zk$, y and z may be determined for an arbitrary value of x from (1.54). For $x = 1$ m, y and z are evaluated as $\frac{2}{13}$ m and $\frac{5}{13}$ m. For $x = -1$ m, y and z are evaluated as $-\frac{9}{26}$ m and $-\frac{11}{39}$ m. The ray coordinates for the line $ \$ $ can now be evaluated from the 2×2 determinants of the array

$$\begin{bmatrix} 1 & 1 & \frac{2}{13} & \frac{5}{13} \\ 1 & -1 & -\frac{9}{26} & -\frac{11}{39} \end{bmatrix} \tag{1.93}$$

and

$$L = \begin{vmatrix} 1 & 1 \\ 1 & -1 \end{vmatrix} = -2, \quad M = \begin{vmatrix} 1 & \frac{2}{13} \\ 1 & -\frac{9}{26} \end{vmatrix} = -\frac{1}{2}, \tag{1.94}$$

$$N = \begin{vmatrix} 1 & \frac{5}{13} \\ 1 & -\frac{11}{39} \end{vmatrix} = -\frac{2}{3}, \quad P = \begin{vmatrix} \frac{2}{13} & \frac{5}{13} \\ -\frac{9}{26} & -\frac{11}{39} \end{vmatrix} = \frac{7}{78} \text{ m},$$

$$Q = \begin{vmatrix} \frac{5}{13} & 1 \\ -\frac{11}{39} & -1 \end{vmatrix} = -\frac{4}{39} \text{ m}, \quad R = \begin{vmatrix} 1 & \frac{2}{13} \\ -1 & -\frac{9}{26} \end{vmatrix} = -\frac{5}{26} \text{ m}.$$

These numerical results satisfy the orthogonality condition of (1.49). The ray coordinates of the line of intersection $ \$ $ may now be written as $\hat{s} = \{S; S_{OL}\}$, where

$$S = -2i - \frac{1}{2}j - \frac{2}{3}k, \tag{1.95}$$

$$S_{OL} = \left(\frac{7}{78}i - \frac{4}{39}j - \frac{5}{26}k \right) \text{ m}.$$

Comparing (1.92) and (1.95), it is apparent that $\hat{s} = \sigma \Delta \hat{S}$, where for this example $\sigma = \frac{1}{78}$.

1.7 The Pencil of Planes through a Line

1.7.1 The Plane Defined by a Line and a Point

The plane containing the line $\{S_1; S_{OL1}\}$ can be rotated about the line, and, in this way, a pencil or single infinity of planes is generated. Imposing the constraint that the plane passes through a point A with position vector r_0 yields a unique plane (see Figure 1.11). The direction vectors S_1, $(r_1 - r_0)$, and $(r - r_0)$, where r_1 is a vector to some point on the line and r is a vector to any point on the plane, are clearly coplanar and, therefore,

$$(r - r_0) \cdot (r_1 - r_0) \times S_1 = 0. \tag{1.96}$$

There is no ambiguity as to the order of operations in the above equation, since a meaningful result occurs only if the cross product is performed prior to the scalar product. Expanding (1.96), regrouping terms, and making the substitution $r_1 \times S_1 = S_{OL1}$ gives

$$r \cdot (S_{OL1} - r_0 \times S_1) - r_0 \cdot (S_{OL1} - r_0 \times S_1) = 0, \tag{1.97}$$

which reduces to

$$r \cdot (S_{OL1} - r_0 \times S_1) - r_0 \cdot S_{OL1} = 0. \tag{1.98}$$

From (1.98), the homogeneous coordinates of the plane are, thus, $[-r_0 \cdot S_{OL1}; S_{OL1} - r_0 \times S_1]$. If point A happened to lie on the line, then $r_0 \times S_1 = S_{OL1}$, and the point and line do not define a unique plane. For this case, substituting $S_{OL1} = r_0 \times S_1$ into (1.98) causes the equation to vanish for all r. When $r_0 = 0$, (1.98) reduces to

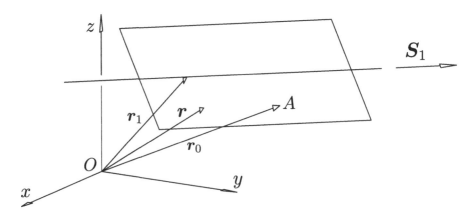

Figure 1.11 Plane determined by a point and a line

$$r \cdot S_{OL1} = 0. \tag{1.99}$$

This simple result is the equation of a plane that passes through the origin and that contains the line $\{S_1; S_{OL1}\}$.

1.7.2 The Plane That Contains a Line and Is Parallel to a Second Line

A unique plane containing the line with coordinates $\{S_1; S_{OL1}\}$ can also be determined by imposing the constraint that the plane be parallel to or contain a second line $\{S_2; S_{OL2}\}$ (see Figure 1.12). In this case, the vector $(r - r_1)$ lies in the plane, and the vector $S_1 \times S_2$ is normal to the plane, where r is a vector to any general point on the plane. It may, therefore, be written that

$$(r - r_1) \cdot (S_1 \times S_2) = 0. \tag{1.100}$$

Expanding (1.100) yields

$$r \cdot (S_1 \times S_2) - r_1 \cdot (S_1 \times S_2) = 0. \tag{1.101}$$

Rearranging the second vector triple product[8] yields

$$r \cdot (S_1 \times S_2) - (r_1 \times S_1) \cdot S_2 = 0. \tag{1.102}$$

Substituting $r_1 \times S_1 = S_{OL1}$ gives

$$r \cdot (S_1 \times S_2) - S_{OL1} \cdot S_2 = 0. \tag{1.103}$$

The homogeneous coordinates for the plane are, therefore, $[-S_{OL1} \cdot S_2; S_1 \times S_2]$.

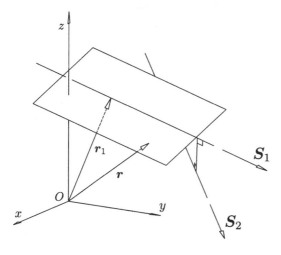

Figure 1.12 The plane through a line and parallel to a second line

---

[8] It can be proven that $a \cdot (b \times c) = (a \times b) \cdot c$. This is also equal to the 3×3 determinant $|abc|$, where a, b, and c are the first, second, and third rows of the determinant, respectively.

A special case needs to be considered. When the two given lines are parallel, $S_1 \times S_2 = 0$, and

$$
\begin{aligned}
S_{OL1} \cdot S_2 &= (r_1 \times S_1) \cdot S_2 \\
&= r_1 \cdot (S_1 \times S_2) \\
&= 0.
\end{aligned}
$$

Thus, (1.103) vanishes identically for all r. In this case, there is a pencil of planes containing the line $\{S_1; S_{OL1}\}$, each of which is parallel to the line $\{S_2; S_{OL2}\}$. The next section focuses on the unique one of these planes that contains both lines.

1.7.3   The Plane Defined by a Pair of Parallel Lines

The equation for the plane through a pair of parallel lines for which $S_1 \times S_2 = 0$ can be most conveniently determined using the vectors p_1 and p_2, which represent the unique point on each line such that the direction of p_1 is perpendicular to S_1 and p_2 is perpendicular to S_2 (see Figure 1.13). Assuming that S_1 and S_2 are unit vectors, then (1.59) may be used to write p_1 and p_2 as

$$
\begin{aligned}
p_1 &= S_1 \times S_{OL1}, && (1.104) \\
p_2 &= S_2 \times S_{OL2}.
\end{aligned}
$$

The perpendicular distance between the two lines may be written as $|p_2 - p_1|$. Clearly $(r - p_1)$, the direction of $(p_1 - p_2)$, and S_1 are coplanar, where r is any vector from point O to a point on the plane. Also the vector $S_1 \times (p_1 - p_2)$ must be perpendicular to the plane. The equation for the plane is, therefore,

$$
(r - p_1) \cdot (S_1 \times (p_1 - p_2)) = 0. \tag{1.105}
$$

Note that in this equation, the expression $p_1 - p_2$ defines a direction that is perpendicular to the plane. As such, it can be considered to be dimensionless. Rearranging

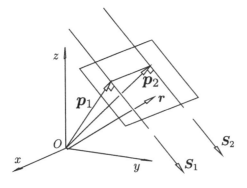

Figure 1.13 The plane through a pair of parallel lines

this equation and substituting $S_1 \times p_1 = -S_{OL1}$ and $S_1 \times p_2 = S_2 \times p_2 = -S_{OL2}$ gives

$$(r - p_1) \cdot (S_{OL2} - S_{OL1}) = 0. \tag{1.106}$$

Substituting $p_1 = S_1 \times S_{OL1}$ and rearranging gives

$$r \cdot (S_{OL2} - S_{OL1}) + (S_1 \times S_{OL1}) \cdot (S_{OL1} - S_{OL2}) = 0, \tag{1.107}$$

which reduces to

$$r \cdot (S_{OL2} - S_{OL1}) - (S_1 \times S_{OL1}) \cdot S_{OL2} = 0. \tag{1.108}$$

The homogeneous coordinates for the plane are, therefore, $[-(S_1 \times S_{OL1}) \cdot S_{OL2}; S_{OL2} - S_{OL1}]$. The coordinates as written should be divided by a unit of length so that the vector part will be dimensionless and the scalar part will have units of length. When $S_{OL1} = S_{OL2}$, the lines are the same and (1.108) vanishes identically.

The perpendicular distance between the two lines can be determined as $|p_2 - p_1|$. The vector $p_2 - p_1$ is given by

$$p_2 - p_1 = S_2 \times S_{OL2} - S_1 \times S_{OL1}. \tag{1.109}$$

The magnitude of this vector, remembering that S_1 and S_2 are unit vectors, is

$$|p_2 - p_1| = |S_{OL2} - S_{OL1}|. \tag{1.110}$$

1.8 A Line and a Plane Determine a Point

The coordinates of a line and a plane are given as $\{S_1; S_{OL1}\}$ and $[D_{O2}; S_2]$, as shown in Figure 1.14, and their equations may be written as

$$r_1 \times S_1 = S_{OL1}, \tag{1.111}$$

$$r_2 \cdot S_2 + D_{O2} = 0, \tag{1.112}$$

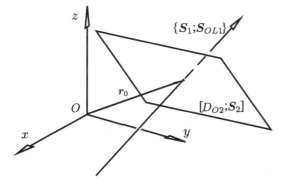

Figure 1.14 A line and a plane determine a point

where r_1 and r_2 are vectors to any point on the line and plane, respectively. The point of intersection of the line and the plane will be designated by the vector r_0, and this vector must satisfy both (1.111) and (1.112) and, therefore,

$$r_0 \times S_1 = S_{OL1}, \tag{1.113}$$

$$r_0 \cdot S_2 + D_{O2} = 0. \tag{1.114}$$

Forming a vector product of S_2 with (1.113) gives

$$S_2 \times (r_0 \times S_1) = S_2 \times S_{OL1}. \tag{1.115}$$

Expanding the left side of (1.115) gives

$$r_0 (S_2 \cdot S_1) - S_1 (S_2 \cdot r_0) = S_2 \times S_{OL1}, \tag{1.116}$$

and substituting (1.114) gives

$$r_0 (S_2 \cdot S_1) = S_2 \times S_{OL1} - D_{O2}S_1 \tag{1.117}$$

and, thus,

$$r_0 = \frac{S_2 \times S_{OL1} - D_{O2}S_1}{S_2 \cdot S_1} \tag{1.118}$$

The homogeneous coordinates of the point of intersection are $(S_2 \cdot S_1; S_2 \times S_{OL1} - D_{O2} S_1)$. When $S_2 \cdot S_1 = 0$, then the line is parallel to the plane and the point of intersection is at infinity with coordinates $\{0; S_1\}$ unless the line lies in the plane and there is no unique point of intersection.

1.9   Determination of the Point on a Line That Is Closest to a Given Point

Figure 1.15 shows a line whose coordinates are given as $\{S ; S_{OL}\}$. The coordinates of a point are given by p_1, and the objective is to find the point on the line that is closest to this point. This closest point is denoted by p_2, and the vector from p_1 to p_2 is shown as d. It is apparent that $d \perp S$.

The point p_2 must satisfy the equation of the line and, thus,

$$p_2 \times S = S_{OL}. \tag{1.119}$$

Since $d \perp S$,

$$d \cdot S = 0. \tag{1.120}$$

It is apparent from 1.15 that

$$p_2 = p_1 + d, \tag{1.121}$$

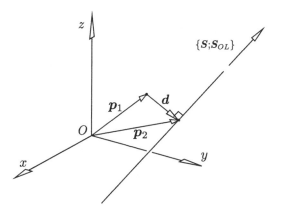

Figure 1.15 Closest point on a line to a given point

and substituting (1.121) into (1.119) gives

$$(\boldsymbol{p}_1 + \boldsymbol{d}) \times \boldsymbol{S} = \boldsymbol{S}_{OL}. \tag{1.122}$$

Expanding (1.122) and rearranging gives

$$\boldsymbol{d} \times \boldsymbol{S} = \boldsymbol{S}_{OL} - \boldsymbol{p}_1 \times \boldsymbol{S}. \tag{1.123}$$

Applying a cross product of $\boldsymbol{S}$ to both sides of (1.123) gives

$$\boldsymbol{S} \times (\boldsymbol{d} \times \boldsymbol{S}) = \boldsymbol{S} \times (\boldsymbol{S}_{OL} - \boldsymbol{p}_1 \times \boldsymbol{S}). \tag{1.124}$$

Expanding the left side of (1.124) gives

$$(\boldsymbol{S} \cdot \boldsymbol{S})\,\boldsymbol{d} - (\boldsymbol{S} \cdot \boldsymbol{d})\,\boldsymbol{S} = \boldsymbol{S} \times (\boldsymbol{S}_{OL} - \boldsymbol{p}_1 \times \boldsymbol{S}). \tag{1.125}$$

Substituting (1.120) into (1.125) gives

$$\boldsymbol{d} = \frac{\boldsymbol{S} \times (\boldsymbol{S}_{OL} - \boldsymbol{p}_1 \times \boldsymbol{S})}{\boldsymbol{S} \cdot \boldsymbol{S}}. \tag{1.126}$$

The coordinates of $\boldsymbol{p}_2$ are determined by substituting (1.126) into (1.121) to yield

$$\boldsymbol{p}_2 = \boldsymbol{p}_1 + \frac{\boldsymbol{S} \times (\boldsymbol{S}_{OL} - \boldsymbol{p}_1 \times \boldsymbol{S})}{\boldsymbol{S} \cdot \boldsymbol{S}}. \tag{1.127}$$

1.10 The Mutual Moment of Two Lines

The Plücker coordinates of two skew lines in space are given as $\{\boldsymbol{S}_1; \boldsymbol{S}_{OL1}\}$ and $\{\boldsymbol{S}_2; \boldsymbol{S}_{OL2}\}$, where $\boldsymbol{S}_1$ and $\boldsymbol{S}_2$ are unit vectors (see Figure 1.16) and their vector equations may be written as

$$\boldsymbol{r}_1 \times \boldsymbol{S}_1 = \boldsymbol{S}_{OL1}, \tag{1.128}$$

$$\boldsymbol{r}_2 \times \boldsymbol{S}_2 = \boldsymbol{S}_{OL2}, \tag{1.129}$$

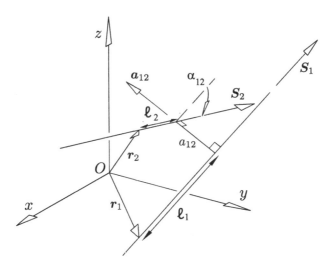

Figure 1.16 A pair of skew lines

where r_1 and r_2 are vectors to any point on the first and second line, respectively. The perpendicular distance between the lines is labeled as a_{12}, and the twist angle between the lines is labeled α_{12} and is measured in a right-hand sense about the vector a_{12}, where a_{12} is a unit vector whose direction is parallel, or anti-parallel, to $S_1 \times S_2$.

The projection of any moment vector $(r_2 - r_1) \times S_2$ on S_1 is the scalar quantity $[(r_2 - r_1) \times S_2] \cdot S_1$, which will be shown to be invariant with the position vectors r_1 and r_2. This quantity, which can also be obtained by projecting the moment vector $(r_1 - r_2) \times S_1$ on S_2, is defined as the mutual moment of two lines. Now the mutual moment may be written as

$$(r_2 - r_1) \times S_2 \cdot S_1 = r_2 \times S_2 \cdot S_1 + r_1 \times S_1 \cdot S_2. \tag{1.130}$$

Substituting (1.128) and (1.129) into the right side of (1.130) yields

$$(r_2 - r_1) \times S_2 \cdot S_1 = S_1 \cdot S_{OL2} + S_2 \cdot S_{OL1}. \tag{1.131}$$

Note that the mutual moment is readily calculated from the coordinates of the two lines and that this quantity has units of length.

The mutual moment will now be calculated in a different manner based on the geometry. From Figure 1.16,

$$(r_2 - r_1) + \ell_2 \, S_2 - a_{12} \, a_{12} - \ell_1 \, S_1 = 0. \tag{1.132}$$

Therefore, the mutual moment may now be calculated as

$$(r_2 - r_1) \times S_2 \cdot S_1 = (\ell_1 \, S_1 + a_{12} \, a_{12} - \ell_2 \, S_2) \times S_2 \cdot S_1. \tag{1.133}$$

Rearranging this equation and recognizing that $S_1 \times S_2 \cdot S_1 = 0$ and $S_2 \times S_2 \cdot S_1 = 0$ gives

$$(r_2 - r_1) \times S_2 \cdot S_1 = a_{12} \, a_{12} \times S_2 \cdot S_1. \tag{1.134}$$

Equation (1.134) may be rewritten as

$$(\boldsymbol{r}_2 - \boldsymbol{r}_1) \times \boldsymbol{S}_2 \cdot \boldsymbol{S}_1 = -a_{12}\,\boldsymbol{a}_{12} \cdot \boldsymbol{S}_1 \times \boldsymbol{S}_2. \tag{1.135}$$

Recall that $\boldsymbol{S}_1$ and $\boldsymbol{S}_2$ are unit vectors and that the unit vector $\boldsymbol{a}_{12}$ is either parallel or anti-parallel to $\boldsymbol{S}_1 \times \boldsymbol{S}_2$. Thus,

$$\boldsymbol{a}_{12} = \pm \frac{\boldsymbol{S}_1 \times \boldsymbol{S}_2}{|\boldsymbol{S}_1 \times \boldsymbol{S}_2|} \tag{1.136}$$

and

$$\boldsymbol{S}_1 \times \boldsymbol{S}_2 = \sin \alpha_{12}\,\boldsymbol{a}_{12}. \tag{1.137}$$

Once the direction for the unit vector $\boldsymbol{a}_{12}$ is chosen, the angle α_{12} is defined as the angle between $\boldsymbol{S}_1$ and $\boldsymbol{S}_2$, measured in a right-hand sense about $\boldsymbol{a}_{12}$. The sine and cosine of this angle may be determined from

$$\boldsymbol{S}_1 \cdot \boldsymbol{S}_2 = \cos \alpha_{12}, \tag{1.138}$$

$$\boldsymbol{S}_1 \times \boldsymbol{S}_2 \cdot \boldsymbol{a}_{12} = \sin \alpha_{12}. \tag{1.139}$$

Substituting (1.139) into (1.135) gives

$$(\boldsymbol{r}_2 - \boldsymbol{r}_1) \times \boldsymbol{S}_2 \cdot \boldsymbol{S}_1 = -a_{12} \sin \alpha_{12}. \tag{1.140}$$

The mutual moment is, thus, a function of the perpendicular distance between the lines and the angle between the directions of the lines and is invariant with the choice of the position vectors $\boldsymbol{r}_1$ and $\boldsymbol{r}_2$.

Equating the right sides of (1.131) and (1.140) gives[9]

$$-a_{12} \sin \alpha_{12} = \boldsymbol{S}_1 \cdot \boldsymbol{S}_{OL2} + \boldsymbol{S}_2 \cdot \boldsymbol{S}_{OL1} \tag{1.141}$$

and further

$$-a_{12} \sin \alpha_{12} = L_1 P_2 + M_1 Q_2 + N_1 R_2 + L_2 P_1 + M_2 Q_1 + N_2 R_1. \tag{1.142}$$

When $a_{12} = 0$, the lines intersect at a finite point, and when $\sin \alpha_{12} = 0$, they intersect at a point at infinity, and in either case the mutual moment of the lines is zero. For the general case, the vector $\boldsymbol{a}_{12}$ is chosen as shown in (1.136). The angle α_{12} may then be determined from (1.138) and (1.139) and then the distance a_{12} is determined from (1.141). A negative value for a_{12} will result if the "direction of travel" along the mutual perpendicular from the first line to the second line is opposite to the direction of the vector $\boldsymbol{a}_{12}$.

As an additional example, consider a line with coordinates $\{L, M, N; P, Q, R\}$. The coordinates of the lines along the x, y, and z axes are $\{1, 0, 0; 0, 0, 0\}$, $\{0, 1, 0; 0, 0, 0\}$, and $\{0, 0, 1; 0, 0, 0\}$, respectively and it can be easily shown that the mutual moment of the line with these three coordinate axes are P, Q, and R, respectively.

---

[9] Recall that during the derivation of this expression the Plücker coordinates of the two lines were written such that $|\boldsymbol{S}_1| = |\boldsymbol{S}_2| = 1$.

1.10.1   Numerical Example

The coordinates of two lines are given as

$$\{S_1; S_{OL1}\} = \{1, 2, 1; -2, 1, 0\}, \tag{1.143}$$

$$\{S_2; S_{OL2}\} = \{-3, 1, 0; 1, 3, 5\}, \tag{1.144}$$

where the direction vectors S_i are dimensionless, and the moment terms S_{OLi} have units of meters, $i = 1, 2$. It is desired to determine the angle between the line directions, α_{12}, and the perpendicular distance between the lines, a_{12}, based on the choice for the direction of the vector $\boldsymbol{a}_{12}$.

The first step will be to scale the line coordinates such that the direction vector is a unit vector. Dividing all terms of the first line by $\sqrt{6}$ and the second line by $\sqrt{10}$ gives

$$\{S_{1u}; S_{OL1u}\} = \{0.4082, 0.8165, 0.4082; -0.8165, 0.4082, 0\}, \tag{1.145}$$

$$\{S_{2u}; S_{OL2u}\} = \{-0.9487, 0.3162, 0; 0.3162, 0.9487, 1.5811\}. \tag{1.146}$$

The direction of the unit vector that is perpendicular to both lines is selected as

$$\boldsymbol{a}_{12} = \frac{S_{1u} \times S_{2u}}{|S_{1u} \times S_{2u}|} = \begin{bmatrix} -0.1302 \\ -0.3906 \\ 0.9113 \end{bmatrix}. \tag{1.147}$$

The angle α_{12} is defined as the angle swept from the direction of the first line to the direction of the second as measured in a right-hand sense about $\boldsymbol{a}_{12}$. The sine and cosine of $\boldsymbol{a}_{12}$ are calculated as

$$\sin \alpha_{12} = (S_{1u} \times S_{2u}) \cdot \boldsymbol{a}_{12} = 0.9916, \tag{1.148}$$

$$\cos \alpha_{12} = S_{1u} \cdot S_{2u} = -0.1291, \tag{1.149}$$

and the angle α_{12} is calculated as

$$\alpha_{12} = 1.700 \text{ radians} = 97.4°. \tag{1.150}$$

The mutual moment of the two lines is calculated as

$$MM = S_{1u} \cdot S_{OL2u} + S_{2u} \cdot S_{OL1u} = 2.4529 \text{ m}. \tag{1.151}$$

From (1.141),

$$-a_{12} \sin \alpha_{12} = 2.4529. \tag{1.152}$$

Solving for a_{12} gives

$$a_{12} = -2.4736 \text{ m}. \tag{1.153}$$

1.11   Determination of the Unique Perpendicular Line to Two Given Lines

In many instances, it is necessary to determine the Plücker coordinates of the line that is mutually perpendicular to two given lines $\{S_1; S_{OL1}\}$ and $\{S_2; S_{OL2}\}$, where

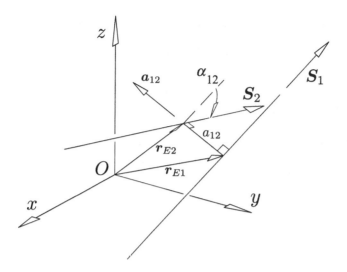

Figure 1.17 Definition of the points r_{E1} and r_{E2}

again for this case $|S_1| = |S_2| = 1$. One method to determine the coordinates of this line will be developed here. A second method, utilizing a concept called the motor product, will be introduced in Section 3.12.

The direction of this line has already been determined as a_{12}, which is a unit vector either parallel or anti-parallel to $\{S_1 \times S_2\}$. Two points on this line are r_{E1} and r_{E2}, which are the intersections of the mutual perpendicular with the two given lines, as shown in Figure 1.17. The determination of either of these points will allow for the determination of the moment of the mutual perpendicular line, which is all that remains to be found in order to define its Plücker coordinates.

Since r_{E1} and r_{E2} lie on the first and second line, respectively,

$$r_{E1} \times S_1 = S_{OL1}, \tag{1.154}$$

$$r_{E2} \times S_2 = S_{OL2}. \tag{1.155}$$

Since $r_{E2} = r_{E1} + a_{12}a_{12}$, (1.155) may be written as

$$(r_{E1} + a_{12}\,a_{12}) \times S_2 = S_{OL2}. \tag{1.156}$$

Expanding the cross product and rearranging terms gives

$$r_{E1} \times S_2 = S_{OL2} - a_{12}\,a_{12} \times S_2. \tag{1.157}$$

Forming the cross product of (1.157) with S_{OL1} yields

$$S_{OL1} \times (r_{E1} \times S_2) = S_{OL1} \times (S_{OL2} - a_{12}\,a_{12} \times S_2), \tag{1.158}$$

and expanding the left side of this equation gives

$$r_{E1}(S_{OL1} \cdot S_2) - S_2(S_{OL1} \cdot r_{E1}) = S_{OL1} \times (S_{OL2} - a_{12}\,a_{12} \times S_2). \tag{1.159}$$

Since r_{E1} lies on the first line, $S_{OL1} \cdot r_{E1} = 0$, and (1.159) can be solved for r_{E1} as

$$r_{E1} = \frac{S_{OL1} \times (S_{OL2} - a_{12}\, a_{12} \times S_2)}{S_{OL1} \cdot S_2}. \tag{1.160}$$

The Plücker coordinates of the line that is mutually perpendicular to the two given lines can, thus, be written as $\{a_{12}; r_{E1} \times a_{12}\}$.

A special case exists for the solution of r_{E1} if $S_{OL1} \cdot S_2 = 0$, which can occur if $S_{OL1} = 0$, if $S_2 = 0$, or if S_{OL1} is perpendicular to S_2. If $S_{OL1} = 0$, then the first line passes through the origin, and the point of intersection of the two lines may be written as

$$r_{E1} = r\, S_1. \tag{1.161}$$

Substituting (1.161) into (1.157) gives

$$r\, S_1 \times S_2 = S_{OL2} - a_{12}\, a_{12} \times S_2. \tag{1.162}$$

Substituting (1.137) into (1.162) gives

$$r\, (a_{12} \sin \alpha_{12}) = S_{OL2} - a_{12}\, a_{12} \times S_2. \tag{1.163}$$

Performing a scalar product of both sides of (1.163) with a_{12} and solving for r gives

$$r = \frac{S_{OL2} \cdot a_{12}}{\sin \alpha_{12}}. \tag{1.164}$$

It is apparent from (1.164) that a further special case exists if $\sin \alpha_{12} = 0$, that is, if the directions of the two lines are parallel. In this special case (of a special case), let r_{E1} be the origin and let $r_{E2} = a_{12}\, a_{12} = \frac{S_2 \times S_{OL2}}{S_2 \cdot S_2}$ be the vector from the origin that is perpendicular to the second line.

If $S_2 = 0$, then the second line is at infinity. The mutual moment of a line with a line at infinity will not be addressed here. The third special case occurs if S_{OL1} and S_2 are perpendicular. Substituting $r_{E1} \times S_1 = S_{OL1}$ into the scalar product $S_{OL1} \cdot S_2 = 0$ gives

$$r_{E1} \times S_1 \cdot S_2 = 0. \tag{1.165}$$

This expression may be written as

$$r_{E1} \cdot S_1 \times S_2 = r_{E1} \cdot a_{12} \sin \alpha_{12} = 0. \tag{1.166}$$

From this last expression it is apparent that the vector r_{E1} is perpendicular to a_{12} and, as such, the vector r_{E1} may be written as a linear combination of the vectors S_1 and S_2, since both of these vectors are also perpendicular to a_{12}. Thus, r_{E1} may be written as

$$r_{E1} = r_1\, S_1 + r_2\, S_2. \tag{1.167}$$

Since r_{E1} lies on the first line,

$$r_{E1} \times S_1 = S_{OL1}. \tag{1.168}$$

Substituting (1.167) into (1.168) gives

$$(r_1 \, S_1 + r_2 \, S_2) \times S_1 = S_{OL1}, \tag{1.169}$$

which simplifies to

$$r_2 \, S_2 \times S_1 = S_{OL1}. \tag{1.170}$$

Substituting (1.137) into (1.170) gives

$$- r_2 \, a_{12} \sin \alpha_{12} = S_{OL1}. \tag{1.171}$$

Performing a scalar product of both sides of (1.171) with a_{12} and solving for r_2 gives

$$r_2 = \frac{-S_{OL1} \cdot a_{12}}{\sin \alpha_{12}}. \tag{1.172}$$

The value for r_1 can be obtained by substituting (1.167) into (1.157). The resulting expression for r_1 is written as

$$r_1 = \frac{S_{OL2} \cdot a_{12}}{\sin \alpha_{12}}. \tag{1.173}$$

1.12   A Pair of Intersecting Lines

Two intersecting lines $\{S_1; S_{OL1}\}$ and $\{S_2; S_{OL2}\}$ are shown in Figure 1.18, where it is assumed that S_1 and S_2 are unit vectors. Two approaches to determine the intersection point of these two given lines will now be presented.

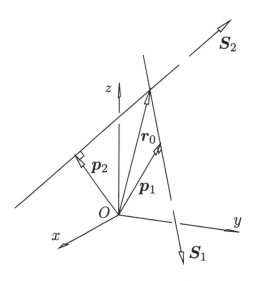

Figure 1.18 A pair of intersecting lines

1.12.1   Approach 1

The vector equations for these two lines are

$$r_1 \times S_1 = S_{OL1}, \tag{1.174}$$

$$r_2 \times S_2 = S_{OL2}, \tag{1.175}$$

where r_1 and r_2 are vectors to any point on the first and second line, respectively. The position vector r_0 indicates the point of intersection of the two lines and, as such, it must satisfy both equations for the lines as

$$r_0 \times S_1 = S_{OL1}, \tag{1.176}$$

$$r_0 \times S_2 = S_{OL2}. \tag{1.177}$$

The vector r_0 will be determined by first forming a cross product of (1.176) with S_2, which yields

$$S_2 \times (r_0 \times S_1) = S_2 \times S_{OL1}. \tag{1.178}$$

Expanding the left side of (1.178) gives

$$r_0(S_2 \cdot S_1) - S_1(r_0 \cdot S_2) = S_2 \times S_{OL1}. \tag{1.179}$$

Forming a scalar product of (1.179) with S_1 yields

$$(r_0 \cdot S_1)(S_2 \cdot S_1) - (S_1 \cdot S_1)(r_0 \cdot S_2) = S_1 \cdot (S_2 \times S_{OL1}). \tag{1.180}$$

Substituting $S_1 \cdot S_1 = 1$ and rearranging gives

$$r_0 \cdot [S_2 - (S_2 \cdot S_1)S_1] = (S_1 \times S_{OL1}) \cdot S_2. \tag{1.181}$$

It remains to solve (1.177) and (1.181) for r_0. Forming the cross product of (1.177) with $[S_2 - (S_2 \cdot S_1)S_1]$ yields

$$[S_2 - (S_2 \cdot S_1)S_1] \times [r_0 \times S_2] = [S_2 - (S_2 \cdot S_1)S_1] \times S_{OL2}. \tag{1.182}$$

Expanding the left side of (1.182) and substituting (1.181) gives

$$r_0\{[S_2 - (S_2 \cdot S_1)S_1] \cdot S_2\} - S_2\{(S_1 \times S_{OL1}) \cdot S_2\} = [S_2 - (S_2 \cdot S_1)S_1] \times S_{OL2}. \tag{1.183}$$

Rearranging this equation gives

$$r_0[1 - (S_1 \cdot S_2)^2] = S_2 \times S_{OL2} - (S_1 \cdot S_2)S_1 \times S_{OL2} + (S_1 \times S_{OL1} \cdot S_2)S_2, \tag{1.184}$$

and solving for r_0 gives

$$r_0 = \frac{S_2 \times S_{OL2} - (S_1 \cdot S_2)S_1 \times S_{OL2} + (S_1 \times S_{OL1} \cdot S_2)S_2}{1 - (S_1 \cdot S_2)^2}. \tag{1.185}$$

The homogeneous coordinates of the point of intersection of the pair of lines are, thus, $(1 - (S_1 \cdot S_2)^2; S_2 \times S_{OL2} - (S_1 \cdot S_2)S_1 \times S_{OL2} + (S_1 \times S_{OL1} \cdot S_2)S_2)$.

1.12.2 Approach 2

A second solution approach to this problem was developed by Dr. David Dooner at the University of Puerto Rico, Mayaguez. Performing a cross product of the left and right sides of (1.176) and (1.177) yields

$$(r_0 \times S_1) \times (r_0 \times S_2) = S_{OL1} \times S_{OL2}. \tag{1.186}$$

Expanding this expression gives

$$[(r_0 \times S_1) \cdot S_2]r_0 - [(r_0 \times S_1) \cdot r_0]S_2 = S_{OL1} \times S_{OL2}. \tag{1.187}$$

Substituting $[(r_0 \times S_1) \cdot r_0] = 0$ and $r_0 \times S_1 = S_{OL1}$ and then solving for r_0 gives

$$r_0 = \frac{S_{OL1} \times S_{OL2}}{S_{OL1} \cdot S_2}. \tag{1.188}$$

The result for the intersection point obtained in (1.188) is much simpler than that in (1.185). However, for the cases where the first line passes through the origin ($S_{OL1} = 0$) or where the two lines lie in a plane that passes through the origin (S_1 and S_2 lie in the plane and, thus, S_{OL1} and S_{OL2} are parallel and perpendicular to the plane), equation (1.188) will reduce to an indeterminate state of $\frac{0}{0}$. Equation (1.185) will yield the correct intersection point for these cases.

1.13 Summary

This chapter introduced representations and notation for points, lines, and planes. It was shown that a point is the dual of a plane and that a line is dual with itself or self-dual. Several geometric problems were solved, such as determining the coordinates of the line that is the intersection of two non-parallel planes. The concept of the mutual moment of two lines was introduced, and a geometric interpretation was presented. Lastly, it was shown how to determine the Plücker coordinates of the line that intersects and is perpendicular to two given lines.

The material presented in this chapter should give the user a firm background in the geometry of points, lines, and planes. Subsequent chapters will expand upon this material to develop an analytic approach for solving velocity, acceleration, and static force balance problems for serial and planar spatial manipulators and mechanisms.

1.14 Problems

1. The equation of a plane is $3x - 4y - 12z - 1 = 0$. Here, the coefficients 3, 4, and 12 are dimensionless, while the coefficient 1 has units of meters. Determine:
   (a) The direction cosines of the unit vector normal to the plane.
   (b) The perpendicular distance of the plane from the origin.

(c) The equation of the plane parallel to the given plane that passes through the origin.

2. Draw the lines with the following Plücker coordinates and determine the perpendicular distance of each line from the origin. The first three coordinates are dimensionless, and the last three coordinates have units of meters.

(a) $\{-4, 12, 3; -24, -6, -8\}$
(b) $\{3, 4, -12; 16, 27, 13\}$
(c) $\{2, 4, -1; -7, 4, 2\}$
(d) $\{4, 1, 5; 5, 15, -7\}$

3. The equations of two planes are

$$-4x + 12y + 3z + 1 = 0,$$
$$3x + 4y - 12z + 1 = 0.$$

Here, the coefficients multiplying the terms x, y, and z are dimensionless, and the remaining coefficient has units of meters. Determine the angle between the planes and the Plücker coordinates of the line of intersection. What is the condition that two planes are perpendicular?

4. Derive the equation for the plane that contains the line $r \times S = S_{OL}$ and the origin where $S = (-4, 12, 3)$ and $S_{OL} = (-24, -6, -8)$. The vector S is dimensionless, and S_{OL} has units of meters.

5. Determine the point of intersection of the line $\{3, 4, -12; 16, 27, 13\}$ and the plane $4x + y + 5z + 1 = 0$. The first three components of the line coordinates are dimensionless, and the last three have units of meters. The coefficients of the equation of the plane that multiply the terms x, y, and z are dimensionless, while the remaining coefficient has units of meters.

6. Find the shortest distance between the straight lines AB and CD when the coordinates of points A, B, C, and D are given in meters as follows:

(a) $A(-2, 4, 3)$, $B(2, -8, 0)$, $C(1, -3, 5)$, $D(4, 1, -7)$
(b) $A(2, 3, 1)$, $B(0, -1, 2)$, $C(1, 2, 5)$, $D(-3, 1, 0)$

7. Prove that if the non-parallel lines $\{S_1; S_{OL1}\}$ and $\{S_2; S_{OL2}\}$ are coplanar, then $S_1 \cdot S_{OL2} + S_2 \cdot S_{OL1} = 0$. Show that they lie in the plane $[S_{OL1} \cdot S_2; S_1 \times S_2]$ and that they intersect at the point $(S_{OL1} \cdot S_2; S_{OL1} \times S_{OL2})$ provided $S_{OL1} \cdot S_2 \neq 0$.

8. A pair of lines $\{S_1; S_{OL1}\}$ and $\{S_2; S_{OL2}\}$ intersect at right angles, where S_1 and S_2 are unit vectors. Derive an equation for the plane that contains the first line and that is perpendicular to the second line. Hence, obtain the expression $r = p_2 + (p_1 \cdot S_2)S_2$ for the position vector of the point of intersection of the lines. Verify this result by simple projection and also deduce that $r = p_1 + (p_2 \cdot S_1)S_1$.

9. Show that the lines AB and CD are coplanar, and find their point of intersection. The coordinates of the four points are given in units of meters as $A(-2, -3, 4)$, $B(2, 3, 0)$, $C(-2, 3, 2)$, $D(2, 0, 1)$. Determine the angle between the lines.

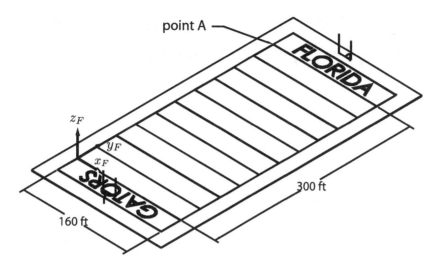

Figure 1.19 Football stadium

10. (a) Show that the three points $(w_1; S_1)$, $(w_2; S_2)$, and $(w_3; S_3)$ determine the plane $[S_1 \cdot S_2 \times S_3; w_1 S_2 \times S_3 + w_2 S_3 \times S_1 + w_3 S_1 \times S_2]$ provided $(S_1 \cdot S_2 \times S_3) \neq 0$, i.e., the points are not linearly dependent (colinear). Assume $w_i \neq 0$, $i = 1 \ldots 3$. (b) Show that the three planes $[D_1; S_1]$, $[D_2; S_2]$, and $[D_3; S_3]$ meet in the point $(S_1 \cdot S_2 \times S_3; D_1 S_2 \times S_3 + D_2 S_3 \times S_1 + D_3 S_1 \times S_2)$ provided $(S_1 \cdot S_2 \times S_3) \neq 0$.

11. Show that the equation of the plane through the origin that contains the line $\{S_1; S_{OL1}\}$ can be expressed in the form $P_1 x + Q_1 y + R_1 z = 0$, where P_1, Q_1, R_1 are the components of the vector S_{OL1}, and x, y, z are the components of S_1.

12. Show that the equation of the plane that contains the line $\{S_1; S_{OL1}\} = \{L_1, M_1, N_1; P_1, Q_1, R_1\}$ and that is parallel to the x axis can be written as $N_1 y - M_1 z - P_1 = 0$.

13. A television camera is located within a stadium (see Figure 1.19). The objective is to determine the position of the camera as measured in terms of the coordinate system shown in the figure.

    The camera is aimed at point A and the unit direction vector from the camera to point A is measured as $[0.45339, 0.84633, -0.27959]^T$. The camera is then pointed at the origin of the reference coordinate system, and the unit direction vector from the camera to the origin is measured as $[0.29892, -0.93611, -0.18533]^T$.

    (a) Determine the Plücker coordinates of the line from the camera to point A and the line from the camera to the origin point.
    (b) Determine the perpendicular distance between the lines.
    (c) If the lines intersect, determine the point of intersection. If the lines do not intersect, determine the midpoint of the line segment that is perpendicular to the two lines.

2 Coordinate Transformations and Manipulator Kinematics

... there occurs to us a double construction of space. In the first construction, we imagine space to be traversed by lines themselves consisting of points. ... In the second construction, these lines are determined by means of planes passing through them. ... A right line of the first description, we shall distinguish by the name of *ray*. ... A right line of the second description, we shall distinguish by the name of *axis*.

Plücker (1865)
"A new Geometry of Space"

2.1 Introduction

This chapter addresses two important concepts. The first is how to define the relative position and orientation of two coordinate systems. With this definition, it is shown how to transform the coordinates of points, lines, and planes from one coordinate system to another. The second objective is to precisely define a kinematic link of a robot manipulator and the types of joints that can interconnect these links. A precise definition is needed so that there is no ambiguity with regards to how the link and joint parameters are measured.

2.2 Relative Pose of Two Coordinate Systems

Figure 2.1 shows the pair of coordinate systems A and B. The pose (position and orientation) of system B relative to system A can be described by six independent parameters. Three of these parameters are the coordinates of the vector $V_{Ao \to Bo}$ that, when expressed in terms of the A coordinate system, represent the position of the origin point of the B coordinate system as seen with respect to the A system. The relative orientation of the B coordinate system can be described by the three angles α, β, and γ. For example, the position and orientation could be defined by imagining

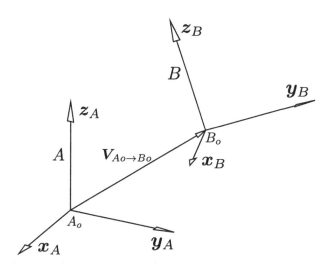

Figure 2.1 Two coordinate systems

the B coordinate system initially aligned with the A system. It is then translated to the point defined by $V_{Ao \rightarrow Bo}$ and then rotated by the angle α about its x axis, by β about its modified y axis, and then by γ about its now twice-modified z axis.

Alternatively, the position and orientation of the B coordinate system measured with respect to the A system can be defined by the vectors $^A V_{Ao \rightarrow Bo}$, $^A x_B$, $^A y_B$, $^A z_B$, where x_B, y_B, and z_B are unit vectors along the coordinate axes of the B coordinate system. The superscript A indicates that the vectors are written in terms of the A coordinate system. From here on, the notation $^I P_J$ will be used to represent the coordinates of a point J as measured with respect to a coordinate system I. As such, $^I P_J$ is a vector that originates at the origin of the I coordinate system and ends at point J and is thus equivalent to $^I V_{Io \rightarrow J}$. Thus, the vector $^A V_{Ao \rightarrow Bo}$ can also be expressed as $^A P_{Bo}$.

The 4 vectors $^A P_{Bo}$, $^A x_B$, $^A y_B$, and $^A z_B$, each of which has 3 scalar components, represent a total of 12 scalar quantities. However, it was previously shown that the relative position and orientation can be defined by six quantities. The explanation for this is that the components of the three unit vectors $^A x_B$, $^A y_B$, and $^A z_B$ are not independent. Since the orientation vectors are unit vectors and are also mutually perpendicular, they must satisfy the following six constraint equations:

$$|^A x_B| = 1, \tag{2.1}$$
$$|^A y_B| = 1,$$
$$|^A z_B| = 1,$$
$$^A x_B \cdot {}^A y_B = 0,$$
$$^A x_B \cdot {}^A z_B = 0,$$
$$^A x_{yB} \cdot {}^A y_B = 0.$$

The unit vectors ${}^A x_B, {}^A y_B, {}^A z_B$ thus represent $9 - 6 = 3$ independent scalar quantities that specify the orientation of coordinate system B relative to A.

Consider now that the coordinate system B is attached to a rigid body. The vectors ${}^A P_{Bo}, {}^A x_B, {}^A y_B$, and ${}^A z_B$, which define the position and orientation of the B coordinate system with respect to the A system and which consist of six independent parameters, can be used to locate the rigid body in space with respect to the A reference frame. Since six independent parameters must be specified to define the position and orientation, it is said that a rigid body in space possesses six degrees of freedom.

2.3   Transformations of Points

In many kinematic problems, the position of a point is known in terms of one coordinate system, and it is necessary to determine the position of the same point measured in another coordinate system. The problem statement is presented as follows (see Figure 2.2):

Given

- ${}^B P_1$, the coordinates of point 1 measured in the B coordinate system, i.e., ${}^B V_{Bo \to 1}$,
- ${}^A P_{Bo}$, the location of the origin of the B coordinate system, measured with respect to the A coordinate system i.e., ${}^A V_{Ao \to Bo}$,
- ${}^A x_B, {}^A y_B$, and ${}^A z_B$, the orientation of the B coordinate system measured with respect to the A coordinate system,

find:

- ${}^A P_1$, the coordinates of point 1 measured in the A coordinate system, i.e., ${}^A V_{Ao \to 1}$.

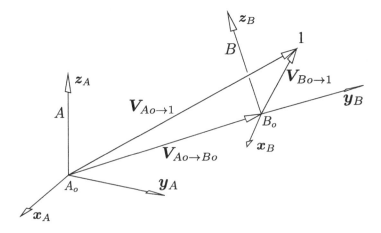

Figure 2.2 Depiction of point transformation problem

From triangle A_0-B_0-1 in Figure 2.2, it may be written that

$$V_{Ao\rightarrow 1} = V_{Ao\rightarrow Bo} + V_{Bo\rightarrow 1}. \qquad (2.2)$$

Evaluating all the vectors in terms of the A coordinate system gives

$${}^A V_{Ao\rightarrow 1} = {}^A V_{Ao\rightarrow Bo} + {}^A V_{Bo\rightarrow 1}. \qquad (2.3)$$

It is thus necessary to solve (2.3) for ${}^A V_{Ao\rightarrow 1}$ ($= {}^A P_1$). The first term on the right side of (2.3) is a given quantity, and the second term is yet to be obtained.

The vector $V_{Bo\rightarrow 1}$ is given in terms of the B coordinate system, and it may be written as

$${}^B V_{Bo\rightarrow 1} = \begin{bmatrix} b_1 \\ b_2 \\ b_3 \end{bmatrix}, \qquad (2.4)$$

where the scalars b_1, b_2, and b_3 are given quantities. This vector can also be written as

$${}^B V_{Bo\rightarrow 1} = b_1 \, {}^B x_B + b_2 \, {}^B y_B + b_3 \, {}^B z_B \qquad (2.5)$$

since it is obvious that ${}^B x_B = [1,0,0]^T$, ${}^B y_B = [0,1,0]^T$, and ${}^B z_B = [0,0,1]^T$. The vectors in (2.5) can all be expressed in terms of the A coordinate system, and thus

$${}^A V_{Bo\rightarrow 1} = b_1 \, {}^A x_B + b_2 \, {}^A y_B + b_3 \, {}^A z_B. \qquad (2.6)$$

All terms on the right side of (2.6) are given quantities. Substituting (2.6) into (2.3) gives

$${}^A V_{Ao\rightarrow 1} = {}^A V_{Ao\rightarrow Bo} + b_1 \, {}^A x_B + b_2 \, {}^A y_B + b_3 \, {}^A z_B, \qquad (2.7)$$

which may be arranged in matrix format as

$${}^A V_{Ao\rightarrow 1} = {}^A V_{Ao\rightarrow Bo} + \begin{bmatrix} {}^A x_B & {}^A y_B & {}^A z_B \end{bmatrix} \begin{bmatrix} b_1 \\ b_2 \\ b_3 \end{bmatrix}. \qquad (2.8)$$

The expression $\begin{bmatrix} {}^A x_B & {}^A y_B & {}^A z_B \end{bmatrix}$ represents a 3×3 matrix that will be designated as

$${}^A_B R = \begin{bmatrix} {}^A x_B & {}^A y_B & {}^A z_B \end{bmatrix}. \qquad (2.9)$$

Substituting ${}^B P_1 = [b_1, b_2, b_3]^T$, ${}^A P_1 = {}^A V_{Ao\rightarrow 1}$, and ${}^A P_{Bo} = {}^A V_{Ao\rightarrow Bo}$ and (2.9) into (2.8) yields

$${}^A P_1 = {}^A P_{Bo} + {}^A_B R \, {}^B P_1. \qquad (2.10)$$

It can be readily shown that (2.10) can be written as

$$\begin{bmatrix} {}^A P_1 \\ 1 \end{bmatrix} = \begin{bmatrix} {}^A_B R & {}^A P_{Bo} \\ 0 \quad 0 \quad 0 & 1 \end{bmatrix} \begin{bmatrix} {}^B P_1 \\ 1 \end{bmatrix}, \qquad (2.11)$$

where the matrix ${}^A_B R$ and the vector ${}^A P_{Bo}$ form the first three rows of a 4×4 matrix. The notation ${}^A_B T$ will be used to represent the 4×4 matrix as

$$\begin{array}{cc} {}^A_B T = \begin{bmatrix} {}^A_B R & {}^A P_{Bo} \\ 0 \quad 0 \quad 0 & 1 \end{bmatrix}. \end{array} \tag{2.12}$$

The point transformation problem can now be written as

$$ {}^A P_1 = {}^A_B T \; {}^B P_1, \tag{2.13}$$

where, from this point on, a vector may be either a three-dimensional vector or a homogeneous coordinate vector, $[x, y, z, w]^T$, depending on the context of the equation it is being used in. It should be apparent that the vectors in (2.10) are three dimensional, while the vectors in (2.13) are written in homogeneous coordinates with $w = 1$.

2.4   Inverse of a Transform

Quite often during robot analyses, it will be necessary to obtain the inverse of a 4×4 transformation. In other words, given ${}^A_B T$ it will be necessary to obtain ${}^B_A T$. The definition of ${}^A_B T$ was presented in (2.12). The matrix ${}^A_B R$ is a 3×3 matrix whose columns are ${}^A x_B$, ${}^A y_B$, ${}^A z_B$, i.e., the coordinates of the unit axis vectors of the B coordinate system measured in the A coordinate system. The vector ${}^A P_{Bo}$ represents the coordinates of the origin of the B coordinate system measured with respect to the A coordinate system.

It should be clear from (2.12) that the inverse ${}^B_A T$ can be written in the form

$$ {}^B_A T = \begin{bmatrix} {}^B_A R & {}^B P_{Ao} \\ 0 \quad 0 \quad 0 & 1 \end{bmatrix}. \tag{2.14}$$

The inverse will be defined once the matrix ${}^B_A R$ and the coordinates of the point ${}^B P_{Ao}$ are determined.

The matrix ${}^A_B R$ can be written in the form

$$ {}^A_B R = \begin{bmatrix} {}^A x_B \cdot {}^A x_A & {}^A y_B \cdot {}^A x_A & {}^A z_B \cdot {}^A x_A \\ {}^A x_B \cdot {}^A y_A & {}^A y_B \cdot {}^A y_A & {}^A z_B \cdot {}^A y_A \\ {}^A x_B \cdot {}^A z_A & {}^A y_B \cdot {}^A z_A & {}^A z_B \cdot {}^A z_A \end{bmatrix}. \tag{2.15}$$

Each of the nine scalar terms of the 3×3 matrix ${}^A_B R$ has been expressed in terms of a scalar product. A scalar product is an invariant operator that can physically be interpreted as being the cosine of the angle between the two unit vectors. The value of the scalar product will remain constant no matter what coordinate system the two vectors are expressed in. Thus, for example,

$$ {}^A x_B \cdot {}^A x_A = {}^B x_B \cdot {}^B x_A. \tag{2.16}$$

Applying this to all the terms of $^A_B R$ yields

$$
^A_B R = \begin{bmatrix} ^B x_B \cdot {}^B x_A & ^B y_B \cdot {}^B x_A & ^B z_B \cdot {}^B x_A \\ ^B x_B \cdot {}^B y_A & ^B y_B \cdot {}^B y_A & ^B z_B \cdot {}^B y_A \\ ^B x_B \cdot {}^B z_A & ^B y_B \cdot {}^B z_A & ^B z_B \cdot {}^B z_A \end{bmatrix}.
\tag{2.17}
$$

It can be seen that the rows of the 3×3 matrix $^A_B R$ are $^B x_A$, $^B y_A$, $^B z_A$ by recognizing that $^B x_B = [1, 0, 0]^T$, $^B y_B = [0, 1, 0]^T$, and $^B z_B = [0, 0, 1]^T$. Thus $^A_B R$ can be written as

$$
^A_B R = \begin{bmatrix} ^A x_B & ^A y_B & ^A z_B \end{bmatrix} = \begin{bmatrix} ^B x_A{}^T \\ ^B y_A{}^T \\ ^B z_A{}^T \end{bmatrix}.
\tag{2.18}
$$

The transpose of (2.18) is

$$
^A_B R^T = \begin{bmatrix} ^B x_A & ^B y_A & ^B z_A \end{bmatrix}.
\tag{2.19}
$$

The columns of the 3×3 matrix in (2.19) are the unit vectors of the A coordinate system measured in terms of the B coordinate system. This is precisely the definition of $^B_A R$, and thus it may be concluded that

$$
^B_A R = {}^A_B R^T.
\tag{2.20}
$$

The remaining term to be determined is $^B P_{Ao}$. This term can readily be calculated from $^A P_{Bo}$ now that $^B_A R$ is known. First, the vector $^A P_{Bo}$ will be transformed to the B coordinate system by utilizing (2.10) as

$$
^B P_{Bo} = {}^B P_{Ao} + {}^B_A R \, {}^A P_{Bo}.
\tag{2.21}
$$

Now $^B P_{Bo} = [0, 0, 0]^T$, which are the coordinates of the origin of the B coordinate system measured in the B system. Substituting this result into (2.21) and rearranging yields

$$
^B P_{Ao} = - {}^B_A R \, {}^A P_{Bo} = - {}^A_B R^T \, {}^A P_{Bo}.
\tag{2.22}
$$

Substituting (2.20) and (2.22) into (2.14) yields the final result

$$
^B_A T = \begin{bmatrix} ^A_B R^T & -{}^A_B R^T \, {}^A P_{Bo} \\ 0 \quad 0 \quad 0 & 1 \end{bmatrix}.
\tag{2.23}
$$

2.5   Standard Transformations

In many problems, the relationship between the coordinate systems will be defined in terms of rotations about the X, Y, or Z axes. A typical problem statement would be as follows: *Given that coordinate system B is initially aligned with coordinate system A and is then rotated α degrees about the X axis, find $^A_B R$ (often written as $R_{x,\alpha}$).* Figure 2.3 shows the A and B coordinate systems. By projection, it can be seen that

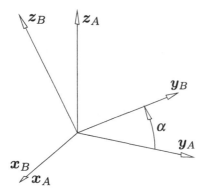

Figure 2.3 Rotation about the X axis

$$^A x_B = \begin{bmatrix} 1 \\ 0 \\ 0 \end{bmatrix},$$

(2.24)

$$^A y_B = \begin{bmatrix} 0 \\ \cos \alpha \\ \sin \alpha \end{bmatrix},$$

(2.25)

$$^A z_B = \begin{bmatrix} 0 \\ -\sin \alpha \\ \cos \alpha \end{bmatrix}.$$

(2.26)

Thus

$$^A_B R = R_{x,\alpha} = \begin{bmatrix} 1 & 0 & 0 \\ 0 & \cos \alpha & -\sin \alpha \\ 0 & \sin \alpha & \cos \alpha \end{bmatrix}.$$

(2.27)

Similarly, for the problem where the B coordinate system is initially aligned with the A system and is then rotated an angle β about the Y axis, the rotation matrix that relates the two coordinate systems can be determined by projection as

$$^A_B R = R_{y,\beta} = \begin{bmatrix} \cos \beta & 0 & \sin \beta \\ 0 & 1 & 0 \\ -\sin \beta & 0 & \cos \beta \end{bmatrix}.$$

(2.28)

Lastly, the B and A coordinate systems are aligned, and the B system is then rotated by an angle γ about the Z axis. Again, the rotation matrix that relates these two coordinate systems, can be found by projection as

$$^A_B R = R_{z,\gamma} = \begin{bmatrix} \cos \gamma & -\sin \gamma & 0 \\ \sin \gamma & \cos \gamma & 0 \\ 0 & 0 & 1 \end{bmatrix}.$$

(2.29)

2.6   General Transformations

Two types of additional problems dealing with transformations can occur. In the first, coordinate system B is initially aligned with coordinate system A. An axis that passes through the origin and an angle of rotation are given, about which coordinate system B will be rotated. The objective is to determine $_B^A R$ for this case. The second problem is the opposite in that a rotation matrix $_B^A R$ is given, and it is desired to determine the axis and angle of rotation that is represented by the matrix. Solutions to both problems will be presented in this section.

2.6.1   Determination of Equivalent Rotation Matrix

In this problem, it is assumed that an axis of rotation, $m = [m_x, m_y, m_z]^T$, and an angle of rotation, θ, are known, where the vector m is a unit vector. A coordinate systems B is initially aligned with a coordinate system A. The B system is then rotated about the axis m, which passes through the origin, by an angle θ (see Figure 2.4). It is desired to find the rotation matrix $_B^A R$.

The problem will be solved by first introducing a coordinate system A' whose Z axis is parallel to the vector m. The relationship between the A and A' coordinate systems can be written as

$$_{A'}^A R = \begin{bmatrix} a_x & b_x & m_x \\ a_y & b_y & m_y \\ a_z & b_z & m_z \end{bmatrix}. \tag{2.30}$$

Only the terms m_x, m_y, and m_z are known in the above equation. The B coordinate system is now initially aligned with the A system, and a new coordinate system, B',

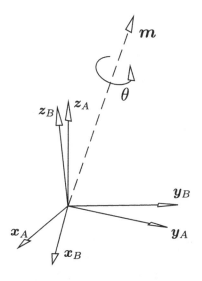

Figure 2.4 Rotation of angle θ about axis m

is aligned with the A' system. The B and B' coordinate systems are now both rotated together about the axis m by the angle θ. Two facts are known. First, the relationship between the B' and the B coordinate systems is identical to the relationship between the A' and the A systems, i.e.,

$$\begin{matrix} B \\ B' \end{matrix} R = \begin{matrix} A \\ A' \end{matrix} R. \tag{2.31}$$

Second, the relationship between the B' and the A' coordinate systems is a simple rotation of θ about the Z axis of the A' coordinate system. Thus

$$\begin{matrix} A' \\ B' \end{matrix} R = \begin{bmatrix} \cos\theta & -\sin\theta & 0 \\ \sin\theta & \cos\theta & 0 \\ 0 & 0 & 1 \end{bmatrix}. \tag{2.32}$$

The rotation matrix that relates the A and B coordinate systems may now be written as

$$\begin{matrix} A \\ B \end{matrix} R = \begin{matrix} A \\ A' \end{matrix} R \begin{matrix} A' \\ B' \end{matrix} R \begin{matrix} B \\ B' \end{matrix} R^T. \tag{2.33}$$

Substituting (2.30), (2.31), and (2.32) into (2.33) yields

$$\begin{matrix} A \\ B \end{matrix} R = \begin{bmatrix} a_x & b_x & m_x \\ a_y & b_y & m_y \\ a_z & b_z & m_z \end{bmatrix} \begin{bmatrix} \cos\theta & -\sin\theta & 0 \\ \sin\theta & \cos\theta & 0 \\ 0 & 0 & 1 \end{bmatrix} \begin{bmatrix} a_x & a_y & a_z \\ b_x & b_y & b_z \\ m_x & m_y & m_z \end{bmatrix}. \tag{2.34}$$

Performing the matrix multiplication and substituting $s = \sin\theta$ and $c = \cos\theta$ yields

$$\begin{matrix} A \\ B \end{matrix} R = \begin{bmatrix} r_{11} & r_{12} & r_{13} \\ r_{21} & r_{22} & r_{23} \\ r_{31} & r_{32} & r_{33} \end{bmatrix}, \tag{2.35}$$

where

$$r_{11} = \cos\theta\left(a_x^2 + b_x^2\right) + m_x^2, \tag{2.36}$$
$$r_{12} = \cos\theta\left(a_y a_x + b_y b_x\right) + \sin\theta\left(a_y b_x - b_y a_x\right) + m_x m_y,$$
$$r_{13} = \cos\theta\left(a_z a_x + b_z b_x\right) + \sin\theta\left(a_z b_x - b_z a_x\right) + m_x m_z,$$
$$r_{21} = \cos\theta\left(a_x a_y + b_x b_y\right) + \sin\theta\left(a_x b_y - b_x a_y\right) + m_x m_y,$$
$$r_{22} = \cos\theta\left(a_y^2 + b_y^2\right) + m_y^2,$$
$$r_{23} = \cos\theta\left(a_z a_y + b_z b_y\right) + \sin\theta\left(a_z b_y - b_z a_y\right) + m_y m_z,$$
$$r_{31} = \cos\theta\left(a_x a_z + b_x b_z\right) + \sin\theta\left(a_x b_z - b_x a_z\right) + m_x m_z,$$
$$r_{32} = \cos\theta\left(a_y a_z + b_y b_z\right) + \sin\theta\left(a_y b_z - b_y a_z\right) + m_y m_z,$$
$$r_{33} = \cos\theta\left(a_z^2 + b_z^2\right) + m_z^2.$$

It must be pointed out that the terms a_x, a_y, a_z, b_x, b_y, and b_z in (2.36) are not known. Three properties of a rotation matrix will be used, however, to resolve this. First, the columns (and rows) of $\begin{matrix} A \\ A' \end{matrix} R$ are unit vectors. Second, the columns (and rows) of $\begin{matrix} A \\ A' \end{matrix} R$ are orthogonal to one another. Third, the last column of $\begin{matrix} A \\ A' \end{matrix} R$ can be calculated as the cross product of the first two columns. These facts will be used to simplify (2.36) and eliminate the unknown terms.

The term r_{11} will be examined first. Since the first row of (2.30) must be a unit vector,

$$a_x^2 + b_x^2 + m_x^2 = 1. \tag{2.37}$$

Substituting $(a_x^2 + b_x^2) = 1 - m_x^2$ into the r_{11} term of (2.36) gives

$$r_{11} = \cos\theta(1 - m_x^2) + m_x^2 = m_x^2(1 - \cos\theta) + \cos\theta, \tag{2.38}$$

and the term r_{11} is now expressed in terms of all known quantities.

The term r_{21} is examined next. Since the first two rows of (2.30) must be orthogonal, it must be the case that

$$a_x a_y + b_x b_y + m_x m_y = 0, \tag{2.39}$$

and thus

$$a_x a_y + b_x b_y = -m_x m_y. \tag{2.40}$$

Further, since the third column of (2.30) can be generated as the cross product of the first two columns, equating the third element of the cross product with the third element of the third column yields

$$a_x b_y - b_x a_y = m_z. \tag{2.41}$$

Substituting (2.40) and (2.41) into the expression for r_{21} in (2.36) gives

$$r_{21} = \cos\theta(-m_x m_y) + \sin\theta(m_z) + m_x m_y. \tag{2.42}$$

Rearranging this equation gives

$$r_{21} = m_x m_y(1 - \cos\theta) + m_z \sin\theta. \tag{2.43}$$

The term r_{21} has now been expressed in terms of all known quantities. Similar substitutions may be made on the remaining elements of $^A_B R$ to eliminate all the unknown terms. The final result for all the terms of the matrix $^A_B R$ is

$$r_{11} = m_x^2(1 - \cos\theta) + \cos\theta, \tag{2.44}$$
$$r_{12} = m_x m_y(1 - \cos\theta) - m_z \sin\theta,$$
$$r_{13} = m_x m_z(1 - \cos\theta) + m_y \sin\theta,$$
$$r_{21} = m_x m_y(1 - \cos\theta) + m_z \sin\theta,$$
$$r_{22} = m_y^2(1 - \cos\theta) + \cos\theta,$$
$$r_{23} = m_y m_z(1 - \cos\theta) - m_x \sin\theta,$$
$$r_{31} = m_x m_z(1 - \cos\theta) - m_y \sin\theta,$$
$$r_{32} = m_y m_z(1 - \cos\theta) + m_x \sin\theta,$$
$$r_{33} = m_z^2(1 - \cos\theta) + \cos\theta.$$

2.6.2   Determination of Axis and Angle of Rotation

For this problem, it is assumed that a rotation matrix, $_B^A R$, is given, and it is desired to calculate the unit axis vector, m, and the angle of rotation about this axis that would rotate coordinate system A so as to align it with coordinate system B. Another way to state this is assuming that coordinate systems B and A were initially aligned, about what axis, m, and by what angle was coordinate system B rotated in order to reach its current orientation?

The rotation matrix may be written as in (2.35), where the numerical values for the terms r_{ij}, $i = 1\ldots3$, $j = 1\ldots3$, are now given. Equation (2.44) has shown how the elements of the rotation matrix can be written in terms of the axis vector m and the rotation angle θ. Equating the summation of the diagonal elements of (2.35) using (2.44) gives

$$r_{11} + r_{22} + r_{33} = (1 - \cos\theta)(m_x^2 + m_y^2 + m_z^2) + 3\cos\theta. \tag{2.45}$$

Since the axis vector m is a unit vector, (2.45) reduces to

$$r_{11} + r_{22} + r_{33} = 1 + 2\cos\theta, \tag{2.46}$$

and solving for $\cos\theta$ gives

$$\cos\theta = \frac{r_{11} + r_{22} + r_{33} - 1}{2}. \tag{2.47}$$

The angle θ is not uniquely defined by (2.47). Two distinct values of θ in the range of $-\pi$ to $+\pi$ exist that will satisfy this equation. The value of θ that lies in the range from 0 to $+\pi$ will be selected, however, and the unique corresponding axis of rotation will be computed.[1]

Now that a value for θ has been determined, the corresponding values for m_x, m_y, and m_z will be obtained. Subtracting the off-diagonal elements of the matrix of equations (2.35) using (2.44) yields

$$r_{21} - r_{12} = 2m_z\sin\theta, \tag{2.48}$$
$$r_{13} - r_{31} = 2m_y\sin\theta,$$
$$r_{32} - r_{23} = 2m_x\sin\theta.$$

The components of the axis vector m can be obtained from (2.48) as

$$m_x = \frac{r_{32} - r_{23}}{2\sin\theta}, \tag{2.49}$$
$$m_y = \frac{r_{13} - r_{31}}{2\sin\theta},$$
$$m_z = \frac{r_{21} - r_{12}}{2\sin\theta}.$$

When the rotation angle is very small, the axis vector m is not well defined as the ratios in (2.49) all approach $\frac{0}{0}$. This is a trivial case, since if the angle of rotation is

[1] Had the value of θ in the range of $-\pi$ to 0 been selected, the resulting rotation axis would point in the opposite direction to the one that was computed when θ was selected in the range of 0 to $+\pi$.

extremely small, the axis of rotation is immaterial. However, when the rotation angle approaches π, the the ratios in (2.49) again approach $\frac{0}{0}$. For this case, an alternate solution for the axis vector $\mathbf{m}$ must be obtained.

Equating the diagonal elements of (2.35) using (2.44) yields

$$r_{11} = m_x^2(1 - \cos\theta) + \cos\theta, \qquad (2.50)$$

$$r_{22} = m_y^2(1 - \cos\theta) + \cos\theta,$$

$$r_{33} = m_z^2(1 - \cos\theta) + \cos\theta.$$

Solving for $m_x, m_y,$ and m_z yields

$$m_x = \pm\sqrt{\frac{r_{11} - \cos\theta}{1 - \cos\theta}}, \qquad (2.51)$$

$$m_y = \pm\sqrt{\frac{r_{22} - \cos\theta}{1 - \cos\theta}},$$

$$m_z = \pm\sqrt{\frac{r_{33} - \cos\theta}{1 - \cos\theta}}.$$

A question is whether the positive or negative solution should be selected for each of the terms $m_x, m_y,$ and m_z. Since the angle θ, as determined from (2.47), was selected to be in the range $0 \leq \theta \leq \pi$, $\sin\theta$ will be in the range $0 \leq \sin\theta \leq 1$. From (2.49), it is apparent that m_x will be positive if the term $r_{32} - r_{23}$ is positive. The sign for the terms m_y and m_z can be deduced in a similar fashion.

Experience has shown that a numerically more accurate determination of $\mathbf{m}$ for the case where θ approaches π results if only the largest magnitude component of $\mathbf{m}$ is calculated from (2.51). The remaining components can be determined from the following equations, which are obtained by summing the off-diagonal elements of the matrix in (2.35) using (2.44):

$$r_{12} + r_{21} = 2m_x m_y(1 - \cos\theta), \qquad (2.52)$$

$$r_{13} + r_{31} = 2m_x m_z(1 - \cos\theta),$$

$$r_{23} + r_{32} = 2m_y m_z(1 - \cos\theta).$$

Thus, for example, if θ is near π and if the absolute value of m_x as calculated in (2.51) is larger than the absolute values of m_y and m_z as calculated from the same equation, then a more accurate solution for m_y and m_z can be obtained from (2.52).

2.7   Transformation of Direction Vectors

Often, in the analysis of robot manipulators, it will be necessary to evaluate a direction vector that is known in one coordinate system in terms of a second coordinate system, where the relationship between the two coordinate systems is given. For example, consider that the relationship between coordinate systems B and A is given by the

4×4 transformation matrix $^A_B T$ and that a direction vector S is known in the B coordinate system (written as $^B S$). It is desired to determine the coordinates of this direction vector in terms of the A coordinate system, i.e., $^A S$.

Previously, it was shown how the coordinates of a *point* that is known in one coordinate system could be determined in a second coordinate system if the relative position and orientation of the two coordinate systems were known. Equations (2.10) and (2.13) gave succinct expressions for this transformation. Now it is desired to transform a *direction vector* rather than a point. A direction vector can be thought of as a dimensionless vector that can be obtained as the difference between two points. The vector is dimensionless in that for *direction*, the concept of length is immaterial. Therefore, the direction vector $^B S$ that is known in terms of the B coordinate system can be defined as the difference between the coordinates of a point along the direction vector, say P_1, and the point at the origin of the B coordinate system, P_{Bo}, and thus may be written as

$$^B S = {}^B P_1 - {}^B P_{Bo}. \tag{2.53}$$

Since $^B P_{Bo} = [0, 0, 0]^T$, the coordinates of point P_1 measured in the B coordinate system may simply be equated to the components of the direction vector $^B S$.

Equation (2.53) may now be expressed in terms of the A coordinate system as

$$^A S = {}^A P_1 - {}^A P_{Bo}. \tag{2.54}$$

From (2.10), this expression may be written as

$$^A S = \left({}^A P_{Bo} + {}^A_B R \, {}^B P_1 \right) - \left({}^A P_{Bo} + {}^A_B R \, {}^B P_{Bo} \right). \tag{2.55}$$

Substituting $^B P_{Bo} = \mathbf{0}$ and simplifying yields

$$^A S = {}^A_B R \, {}^B P_1. \tag{2.56}$$

Since previously the coordinates of the point $^B P_1$, which was any arbitrary point along the direction vector as seen from the B coordinate system, were chosen to equal the components of the direction vector $^B S$, (2.56) may be written as

$$^A S = {}^A_B R \, {}^B S, \tag{2.57}$$

which is the desired result.

2.8   Transformation of Lines

Figure 2.5 shows the line $\{S_1; S_{OL1}\}$ and two reference coordinate systems. It is assumed that the coordinates of the line are known in terms of the B coordinate system, written as $\{{}^B S_1; {}^B S_{OL1}\}$. Also, the relationship between the two coordinate systems is known, and this is written as $^A_B T$. It is desired to determine the coordinates of the line as measured with respect to the A coordinate system, i.e., $\{{}^A S_1; {}^A S_{OL1}\}$.

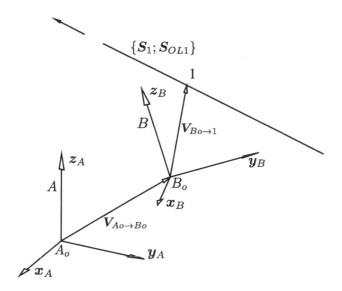

Figure 2.5 Line $\{S_1; S_{OL1}\}$ as seen in two coordinate systems

From the previous section, the direction of the line as measured in the A coordinate system can readily be determined in terms of the given quantities as

$$^{A}S_1 = {}^{A}_{B}R\ {}^{B}S_1, \tag{2.58}$$

where ${}^{A}_{B}R$ is the upper left 3×3 submatrix of ${}^{A}_{B}T$. From Figure 2.5, it is apparent that the moment of the line measured with respect to the A coordinate system may be written as

$$^{A}S_{OL1} = \left({}^{A}V_{Ao \rightarrow Bo} + {}^{A}V_{Bo \rightarrow 1} \right) \times {}^{A}S_1. \tag{2.59}$$

This may be written as

$$^{A}S_{OL1} = \left({}^{A}V_{Ao \rightarrow Bo} \times {}^{A}S_1 \right) + \left({}^{A}V_{Bo \rightarrow 1} \times {}^{A}S_1 \right). \tag{2.60}$$

The term ${}^{A}V_{Ao \rightarrow Bo}$ may be simply written as ${}^{A}P_{Bo}$, i.e., the coordinates of the origin of the B coordinate system as seen with respect to the A coordinate system. These coordinates are the first three terms of the fourth column of ${}^{A}_{B}T$. Thus (2.60) may be written as

$$^{A}S_{OL1} = \left({}^{A}P_{Bo} \times {}^{A}S_1 \right) + \left({}^{A}V_{Bo \rightarrow 1} \times {}^{A}S_1 \right). \tag{2.61}$$

Substituting (2.58) into the left cross product gives

$$^{A}S_{OL1} = \left({}^{A}P_{Bo} \times {}^{A}_{B}R\ {}^{B}S_1 \right) + \left({}^{A}V_{Bo \rightarrow 1} \times {}^{A}S_1 \right). \tag{2.62}$$

The second cross product in (2.60) represents the moment of the line relative to the origin point of the B coordinate system, but this moment vector is evaluated with respect to the A coordinate system. A moment vector is not a line-bound vector (this will be discussed in more detail in Chapter 3). That is, a moment can be thought of as a dimensionless direction vector multiplied by the magnitude of the moment. Writing this moment vector in the B coordinate system and then converting this vector to the A coordinate system allows (2.62) to be written as

$$^{A}S_{OL1} = \left(^{A}P_{Bo} \times {}^{A}_{B}R \; {}^{B}S_{1} \right) + {}^{A}_{B}R \left({}^{B}V_{Bo \rightarrow 1} \times {}^{B}S_{1} \right). \tag{2.63}$$

Lastly, the final cross product term is simply the moment of the line as seen with respect to the B coordinate system, and thus

$$^{A}S_{OL1} = \left(^{A}P_{Bo} \times {}^{A}_{B}R \; {}^{B}S_{1} \right) + {}^{A}_{B}R \; {}^{B}S_{OL1}. \tag{2.64}$$

The Plücker coordinates of the line measured with respect to the A coordinate system may now be written as

$$\{^{A}S_{1}; {}^{A}S_{OL1}\} = \{{}^{A}_{B}R \; {}^{B}S_{1}; {}^{A}P_{Bo} \times {}^{A}_{B}R \; {}^{B}S_{1} + {}^{A}_{B}R \; {}^{B}S_{OL1}\}. \tag{2.65}$$

2.9   Transformations of Planes

Figure 2.6 shows the plane $[D_{O1}; S_{1}]$ and two reference coordinate systems. It is assumed that the coordinates of the plane are known in terms of the B coordinate system, written as $[^{B}D_{O1}; {}^{B}S_{1}]$. Also, the relationship between the two coordinate systems is known, and this is written as ${}^{A}_{B}T$. It is desired to determine the coordinates of the plane as measured with respect to the A coordinate system, i.e., $[^{A}D_{O1}; {}^{A}S_{1}]$.

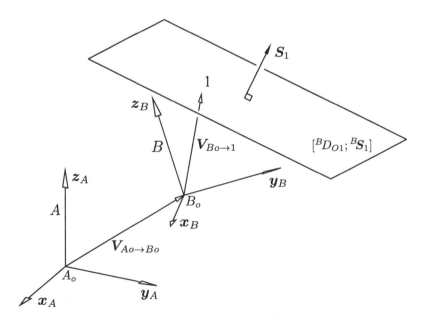

Figure 2.6 Transformation of plane coordinates

The direction of the vector that is perpendicular to the plane as measured in the A coordinate system can readily be determined in terms of the given quantities as

$$^A S_1 = {}^A_B R \, {}^B S_1, \tag{2.66}$$

where ${}^A_B R$ is the upper left 3×3 submatrix of ${}^A_B T$. From Figure 2.6, it is apparent that

$$[^A r - ({}^A V_{Ao \to Bo} + {}^A V_{Bo \to 1})] \cdot {}^A S_1 = 0, \tag{2.67}$$

where $^A r$ is a vector from the origin of the A coordinate system to any point on the plane, and point 1 is a point on the plane. Rearranging this equation yields

$$^A r \cdot {}^A S_1 + (- {}^A V_{Ao \to Bo} - {}^A V_{Bo \to 1}) \cdot {}^A S_1 = 0. \tag{2.68}$$

This equation may be written as

$$^A r \cdot {}^A S_1 + {}^A D_{O1} = 0, \tag{2.69}$$

where

$$^A D_{O1} = (- {}^A V_{Ao \to Bo} - {}^A V_{Bo \to 1}) \cdot {}^A S_1. \tag{2.70}$$

Equation (2.70) can be rearranged as

$$^A D_{O1} = - {}^A V_{Ao \to Bo} \cdot {}^A S_1 - {}^A V_{Bo \to 1} \cdot {}^A S_1. \tag{2.71}$$

The scalar product is invariant with respect to the choice of coordinate system, and thus

$$^A D_{O1} = - {}^A V_{Ao \to Bo} \cdot {}^A S_1 - {}^B V_{Bo \to 1} \cdot {}^B S_1. \tag{2.72}$$

The scalar product $- ({}^B V_{Bo \to 1} \cdot {}^B S_1)$ is equal to the given term $^B D_{O1}$. The term $^A V_{Ao \to Bo}$ may be simply written as $^A P_{Bo}$, i.e., the coordinates of the origin of the B coordinate system as seen with respect to the A coordinate system. These coordinates are the first three terms of the fourth column of ${}^A_B T$. Equation (2.66) is used to substitute for $^A S_1$. This yields

$$^A D_{O1} = - {}^A P_{Bo} \cdot {}^A_B R \, {}^B S_1 + {}^B D_{O1}, \tag{2.73}$$

and the coordinates of the plane with respect to the A coordinate system may be written as

$$[{}^A D_{O1}; {}^A S_1] = [{}^B D_{O1} - {}^A P_{Bo} \cdot {}^A_B R \, {}^B S_1; \, {}^A_B R \, {}^B S_1]. \tag{2.74}$$

2.10   Spatial Links and Joints

A serial chain is formed by a series of rigid body links that are interconnected by joints. In this text, a serial robot manipulator is defined as a kinematic chain where one end is connected to ground while at the free end there is attached an end effector, gripper, or other tooling device. A parallel robot manipulator is defined as an end effector body that is connected to ground by multiple kinematic chains.

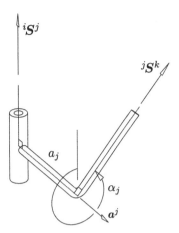

Figure 2.7 Spatial rigid body link

2.10.1   Spatial Link

Figure 2.7 illustrates a link j connecting a pair of consecutive joint axes, labeled by the unit direction vectors $^iS^j$ and $^jS^k$, which are in general skew. Two scalar parameters, the link length a_j and the twist angle α_j, define the relative position and orientation of this pair of skew axes. The link length is the mutual perpendicular distance between the joint axes, and the twist angle is the angle between the joint axes. Specifically, the unit vector a^j is defined as being parallel or antiparallel to $^iS^j \times {}^jS^k$. Once the direction of a^j is chosen, the link length is defined as the distance along the line that is perpendicular to both joint axes as one travels from joint axis $^iS^j$ to $^jS^k$. Note that if the direction of travel is opposite to the direction chosen for a^j, then the link length a_j will be a negative value. The twist angle α_j is measured in a right hand sense about a^j as the angle that is swept from $^iS^j$ to $^jS^k$.

2.10.2   Revolute Joint (R)

The nature of the relative motion between a pair of successive links is determined by the type of connecting joint. One of the simplest and most common joints is the revolute joint, denoted by the letter R. This joint connects two links, as shown in Figure 2.8. Link i is able to rotate relative to link j about the joint axis, whose direction is labeled $^iS^j$.[2] Link j thus has one degree of freedom relative to link i. The joint angle $_i\theta_j$ measures the relative rotation of the two links and is defined as the angle between the unit vectors a^i and a^j, measured in a right hand sense with respect to the unit vector $^iS^j$.

Since link j can only rotate relative to link i, the distance $_iS_j$ is constant. This parameter is called the joint offset distance, and it represents the mutual perpendicular

---

[2] It is assumed that the directions of vector $^iS^j$ of link i and vector $^iS^j$ of link j will always be selected so as to be parallel and not anti-parallel when the joint is assembled.

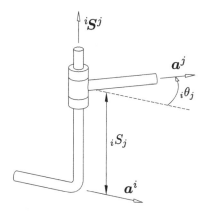

Figure 2.8 Revolute joint

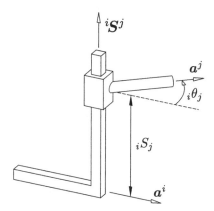

Figure 2.9 Prismatic joint

distance between the link axes a^i and a^j. The joint offset distance will be a positive value if movement from the joint axis a^i to a^j is parallel to the direction of $^iS^j$.

2.10.3 Prismatic Joint (P)

A prismatic joint, which is denoted by the letter P, allows link j to translate parallel to the vector $^iS^j$ with one degree of freedom relative to link i, as shown in Figure 2.9. The angle $_i\theta_j$ is a constant, and it is measured in the same way as for the revolute joint, i.e., it is the angle between the vectors a^i and a^j measured in a right-hand sense about the vector $^iS^j$. The offset distance $_iS_j$ is a variable for the prismatic joint.

2.10.4 Cylindrical Joint (C)

A cylindrical joint, represented by the letter C, allows link j to rotate about and translate parallel to the vector $^iS^j$ relative to link i, as shown in Figure 2.10. Link

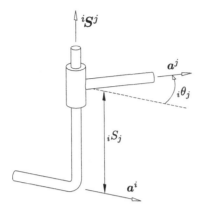

Figure 2.10 Cylindrical joint

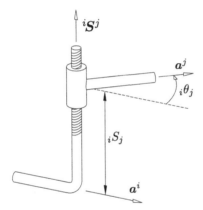

Figure 2.11 Screw joint

j thus has two independent degrees of freedom relative to link i. The joint angle $_i\theta_j$ and the offset distance $_iS_j$ are both variables.

2.10.5   Screw Joint (H)

The screw joint, denoted by the letter H, is shown in Figure 2.11. For this joint, the offset distance $_iS_j$ is related to the joint angle $_i\theta_j$ by the linear equation

$$_iS_j = {_ih_j} \, {_i\theta_j}, \tag{2.75}$$

where $_ih_j$ is the pitch of the screw joint. Clearly, $_ih_j$ is a constant with units of length/radian, and it may be positive or negative according to whether the screw has a right- or left-handed thread. Since the offset distance is a function of the joint angle, link j has one degree of freedom relative to link i.

2.10.6   Higher Order Joints

Three higher order joints, the plane joint denoted by the letter E, the Hooke joint denoted by the letter H, and the spherical or ball-and-socket joint denoted by the letter S, will be presented here. It will be shown how each of these joints can be modeled by a series of revolute and prismatic joints with special geometry for the link lengths, twist angles, joint offsets, and/or joint angles.

The plane joint is illustrated in Figure 2.12. This joint permits three independent degrees of freedom between links i and j. These freedoms can be considered as a pair of linear displacements in the plane of motion together with a rotation about an axis that is perpendicular to the plane of motion. It is not possible to actuate the plane pair in this form in an open loop. However, the plane joint is kinematically equivalent to certain combinations of revolute and prismatic joints. Two such combinations, a PRP and a RRR chain, are shown in Figure 2.13.

For the kinematically equivalent PRP chain, the following special geometry exists:

$$\alpha_i = \frac{\pi}{2}, \;\; a_i = 0, \tag{2.76}$$
$$\alpha_j = \frac{\pi}{2}, \;\; a_j = 0,$$
$$_iS_j = 0, \;\; _h\theta_i = 0, \;\; _j\theta_k = 0.$$

For the kinematically equivalent RRR chain, the following special geometry exists:

$$\alpha_i = 0, \;\; \alpha_j = 0. \tag{2.77}$$

Other combinations of revolute and prismatic joints with appropriate corresponding geometries can be used to form a kinematically equivalent plane joint.

The Hooke joint is simply two revolute joints whose axes intersect. Figure 2.14 shows the case where the axes are perpendicular to one another, but this is not a requirement. It should be apparent that link k possesses two degrees of freedom relative to link i.

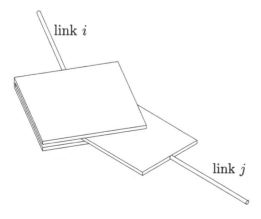

Figure 2.12   Plane joint

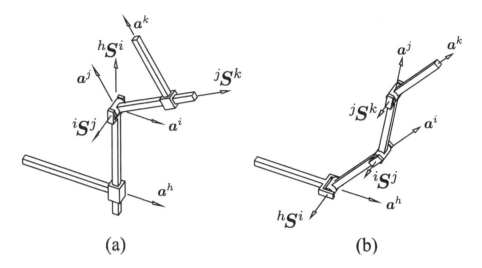

Figure 2.13 Kinematically equivalent planar joint: (a) PRP chain, (b) RRR chain

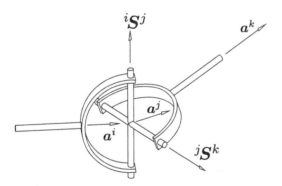

Figure 2.14 Hooke joint

The spherical joint, or ball-and-socket joint, is illustrated in Figure 2.15. Link j has three degrees of freedom relative to link i. These three freedoms can be thought of as the three rotations that will align coordinate system 2, which is attached to link j, with coordination system 1, which is attached to link i, as shown in the figure. The design and implementation of a spherical joint can be a complicated process. It is especially involved if it is necessary to actuate the three freedoms of the joint. The spherical joint, however, can be modeled by three non-coplanar co-intersecting revolute joints, as shown in Figure 2.16. It is not necessary for joint axis vector $^i S^j$ to be perpendicular to joint axis vector $^j S^k$ or for joint axis vector $^j S^k$ to be perpendicular to joint axis vector $^k S^l$, as is shown in the figure. A spherical joint is modeled as long as the three consecutive joint axis vectors intersect at a common point. This method of modeling a spherical joint is commonly used in industrial manipulators.

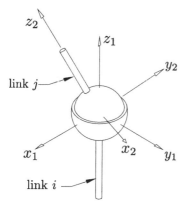

Figure 2.15 Spherical joint

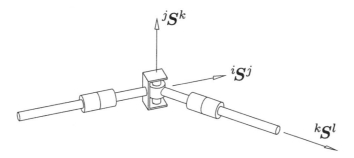

Figure 2.16 Kinematically equivalent spherical joint

2.11 Labeling of a Kinematic Chain

A kinematic chain is shown in Figure 2.17, where the rigid bodies and joint types are labeled starting at 0 (ground) to 6 (distal end of manipulator). The present objective is to:

- select the direction for the joint axis vectors,
- select the directions for the link vectors,
- label the joint angles and twist angles,
- label the offset and link length distances, and
- compile the mechanism parameters in a table listing the constant values and identifying which parameters are variable.

2.11.1 Step 1: Label the Joint Axis Vectors

The first step is to label the joint axes, as is shown in Figure 2.18. As previously stated, the directions of the vectors can arbitrarily point in either direction along the joint axis. However, once the directions are selected, it is important that they be documented for

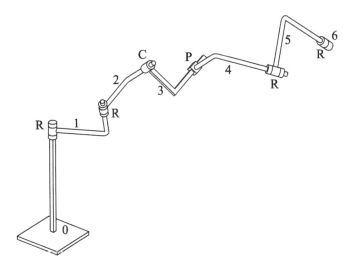

Figure 2.17 Kinematic chain

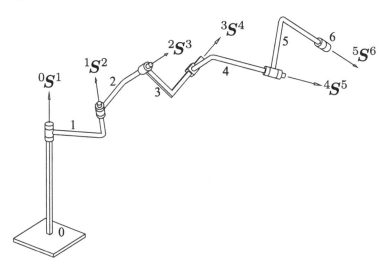

Figure 2.18 Joint vectors labeled

use in all future analyses. The simplest means of labeling the joint axes is to remember that for a revolute, cylindric, or screw joint, the joint axis is along the line of rotation. For a prismatic joint, there is no particular axis since all points in one body undergo the same relative parallel sliding motion. The direction of the joint vector must be parallel to the direction of the translation.

2.11.2   Step 2: Label the Link Vectors

Once the joint axis vectors are specified, the link vectors can be labeled. The link vectors lie along the line that is perpendicular to both of the joint axis vectors that the

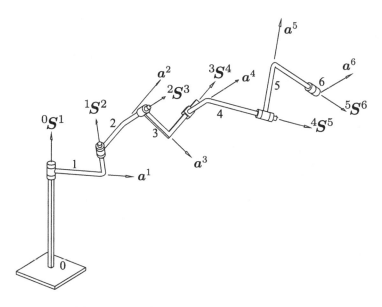

Figure 2.19 Link vectors labeled

link connects, as shown in Figure 2.19. The line perpendicular to two joint axes will be unique unless the two joint axes are parallel. If the two joint axes are parallel, then the location, but not the direction of the link vector, may be arbitrarily selected.

The direction of the last link vector, a^6, which is rigidly attached to link 6, must be perpendicular to $^5S^6$. However, since there is no seventh joint axis, the choice of which direction perpendicular to $^5S^6$ is arbitrary. Also, since there is no seventh joint angle, the line of action of a^6 is not uniquely defined. The point along the sixth joint axis line that the link vector a^6 intersects may be arbitrarily selected. In Figure 2.19, the axis vector was chosen to pass through the point at the center of the end effector tool mounting plate.

2.11.3 Step 3: Label the Joint Angles and Twist Angles

Once the joint axis vectors and the link vectors are specified, the joint angles and twist angles are uniquely defined. For example, as shown in Figure 2.20, $_2\theta_3$ is defined as the angle between a^2 and a^3 measured in a right-hand sense about $^2S^3$, i.e., $a^2 \times a^3 = {}^2S^3 \sin {}_2\theta_3$. Similarly, α_3 is defined is the angle between $^2S^3$ and $^3S^4$ measured in a right-hand sense about a^3, i.e., $^2S^3 \times {}^3S^4 = a^3 \sin \alpha_3$. The joint angles and twist angles are shown in Figures 2.20 and 2.21.

2.11.4 Step 4: Label the Offset and Link Lengths

The offset lengths and link lengths are now uniquely defined. For example, the offset $_2S_3$ is the distance between the lines whose directions are given by a^2 and a^3.

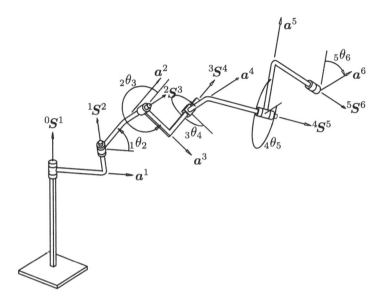

Figure 2.20 Joint angles labeled

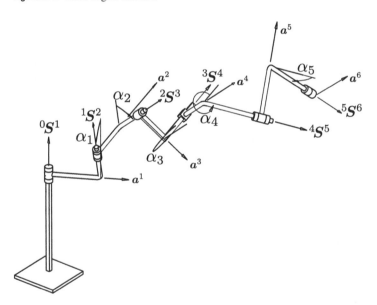

Figure 2.21 Twist angles labeled

Similarly, the link length a_3 is the distance between the lines whose directions are given by $^2S^3$ and $^3S^4$. The offset and link lengths are shown in Figures 2.22 and 2.23.

The offset and link lengths may have negative values. For example, the offset distance $_2S_3$ will be positive if the direction of travel from a^2 to a^3 is along the direction of $^2S^3$. The offset distance $_2S_3$ will be negative if moving from a^2 to a^3 is opposite to the direction of vector $^2S^3$.

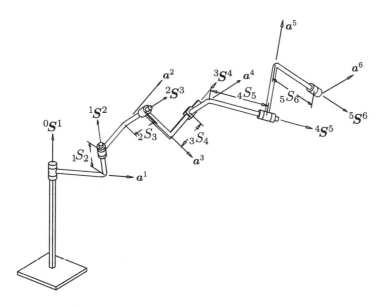

Figure 2.22 Offset Distances Labeled

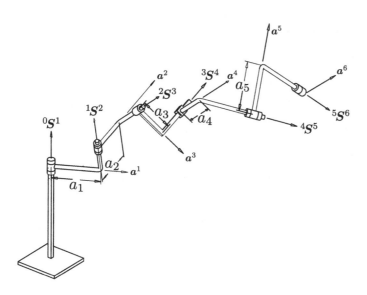

Figure 2.23 Link lengths labeled

It is important to note that the offset distance $_0S_1$ is not defined. According to the labeling convention, $_0S_1$ would be the distance between the link lines whose directions are given by a^0 and a^1. Since a^0 is not defined, the offset distance $_0S_1$ is not defined either.

2.11.5   Step 5: Compilation of Mechanism Parameters

The values for the constant parameters for the kinematic chain must be recorded. The mechanism parameters for the kinematic chain shown in Figure 2.19 are listed in Table 2.1.

Recall that the value for the offset $_5S_6$ was arbitrarily selected by the user, as was the direction for the vector a^6. Although $_5S_6$ looks like a constant parameter in the figures, it is not. The distance $_5S_6$ merely establishes a point on the line along the last joint axis through which the line along the last hypothetical link will pass. For this reason, the parameter $_5S_6$ is not listed in the table as a constant mechanism parameter.

The first joint angle must be measured with respect to ground and not relative to another link as is the case for all the other joint angles. A coordinate system, named the fixed coordinate system, is attached to ground. Its origin is located at the intersection of the lines whose directions are $^0S^1$ and a^1. The Z axis of the fixed coordinate system is along $^0S^1$ (see Figure 2.24). The first joint angle, labeled $_0\phi_1$, is defined as the angle between the X axis of the fixed coordinate system and the line whose direction is specified by a^1, measured in a right-hand sense about $^0S^1$.

Table 2.1. Mechanism parameters for kinematic chain

link length, in	twist angle, deg	joint offset, in	joint angle, deg
$a_1 = 3.25$	$\alpha_1 = 30$		$_0\phi_1 = $ variable
$a_2 = 2.25$	$\alpha_2 = 30$	$_1S_2 = 2.75$	$_1\theta_2 = $ variable
$a_3 = 2.125$	$\alpha_3 = 270$	$_2S_3 = $ variable	$_2\theta_3 = $ variable
$a_4 = 3.5$	$\alpha_4 = 210$	$_3S_4 = $ variable	$_3\theta_4 = 270$
$a_5 = 3.25$	$\alpha_5 = 40$	$_4S_5 = 3.75$	$_4\theta_5 = $ variable
			$_5\theta_6 = $ variable

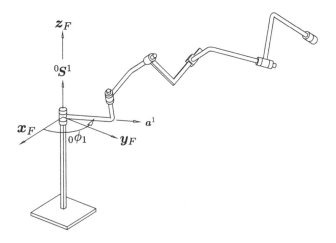

Figure 2.24 Definition of first coordinate system

2.12 Standard Link Coordinate Systems

For the analysis of robot manipulators, it is necessary to attach a coordinate system to each rigid body. The selection of the coordinate system will be done systematically. The coordinate system attached to a link j (see Figure 2.25) will have its origin located at the intersection of $^iS^j$ and a^j. The Z axis of the coordinate system will be parallel to $^iS^j$, and the X axis will be parallel to a^j.

For a serial manipulator, the coordinate system attached to link 1 will be called the first coordinate system. Similarly, the coordinate system attached to link 2 will be called the second coordinate system, and so on. Figure 2.25 shows link i and link j. It is desired to determine the transformation that relates these two coordinate systems.

The jth coordinate system can be obtained by initially aligning it with the ith. It is then translated by the distance a_i along the X axis and rotated by the angle α_i about the X axis. Following this, the coordinate system is translated along the current Z axis by a distance $_iS_j$ and then finally rotated by the angle $_i\theta_j$ about the Z axis to align it with the jth coordinate system. The transformation that relates the i and j coordinate systems can be written as

$$
{}_j^iT =
\begin{bmatrix} 1 & 0 & 0 & a_i \\ 0 & 1 & 0 & 0 \\ 0 & 0 & 1 & 0 \\ 0 & 0 & 0 & 1 \end{bmatrix}
\begin{bmatrix} 1 & 0 & 0 & 0 \\ 0 & c_i & -s_i & 0 \\ 0 & s_i & c_i & 0 \\ 0 & 0 & 0 & 1 \end{bmatrix}
\begin{bmatrix} 1 & 0 & 0 & 0 \\ 0 & 1 & 0 & 0 \\ 0 & 0 & 1 & {}_iS_j \\ 0 & 0 & 0 & 1 \end{bmatrix}
\begin{bmatrix} c_{ij} & -s_{ij} & 0 & 0 \\ s_{ij} & c_{ij} & 0 & 0 \\ 0 & 0 & 1 & 0 \\ 0 & 0 & 0 & 1 \end{bmatrix}
$$
(2.78)

or

$$
{}_j^iT =
\begin{bmatrix} c_{ij} & -s_{ij} & 0 & a_i \\ s_{ij}c_i & c_{ij}c_i & -s_i & -s_i \, {}_iS_j \\ s_{ij}s_i & c_{ij}s_i & c_i & c_i \, {}_iS_j \\ 0 & 0 & 0 & 1 \end{bmatrix},
$$
(2.79)

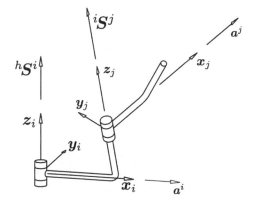

Figure 2.25 Definition of the ith and jth coordinate systems

where s_i and c_i represent the sine and cosine of the twist angle α_i, and s_{ij} and c_{ij} represent the sine and cosine of the joint angle ${}_i\theta_j$.

One additional transformation will be presented for completeness. This is the relationship between the first and the fixed coordinate systems. The transformation is simply a rotation about the Z axis by the angle ϕ because the origins of the fixed and first coordinate systems are coincident. The transformation can be written as

$$
{}_1^F T = \begin{bmatrix} \cos\phi_1 & -\sin\phi_1 & 0 & 0 \\ \sin\phi_1 & \cos\phi_1 & 0 & 0 \\ 0 & 0 & 1 & 0 \\ 0 & 0 & 0 & 1 \end{bmatrix}. \tag{2.80}
$$

By defining a standard method of attaching coordinate systems to each link, it is now possible to obtain the directions of any of the vectors along the links or joint axes in terms of the fixed coordinate system as long as the constant mechanism parameters and values for the variable joint parameters are known. For example, the directions of the vectors ${}^4S^5$ and a^5, measured in the fixed coordinate system, can be obtained as the third and first column respectively of the rotation matrix associated with the transformation ${}_5^F T$. This is because ${}^4S^5$ is along the Z axis, and a^5 is along the X axis of the fifth standard coordinate system. The matrix ${}_5^F T$ can be obtained as

$$
{}_5^F T = {}_1^F T \, {}_2^1 T \, {}_3^2 T \, {}_4^3 T \, {}_5^4 T. \tag{2.81}
$$

This procedure will be used in subsequent chapters to determine the directions of vectors that will be needed in future velocity, acceleration, and static force analyses of serial manipulators.

2.13   Summary

This chapter addressed the problem of how to describe the position and orientation of one coordinate system relative to another. It was shown that a convenient representation for position is the specification of the location of the origin of the second coordinate system relative to the first. Orientation can be defined by specifying the coordinates of the unit axis vectors of the second coordinate system measured in the first coordinate system. It was also shown that the selected method of describing relative position and orientation could be used to easily transform a point between coordinate systems. Homogeneous coordinates were introduced, and the point transformation matrix was expressed as a compact 4×4 matrix whose components had geometric meaning.

Further, in this chapter the rigid body link was defined and quantified by the link length and twist angle. The steps for labeling a kinematic chain and defining the constant mechanism parameters were discussed. Finally, a standard coordinate system was attached to each link, and the transformation between coordinate systems was developed. With knowledge of the variable mechanism parameters, it will now be possible to obtain the coordinates of any point in the kinematic chain and the direction of any vector associated with the chain in terms of a fixed reference frame or any of the individual link coordinate systems.

2.14   Problems

1. A coordinate system B is initially coincident with coordinate system A. It is rotated by an angle θ about the X axis and then subsequently rotated by an angle β about its new Y axis. Determine the orientation relationship of B with respect to A, i.e., $^A_B R$.

2. The coordinates of point 1 as seen from the A coordinate system are $\begin{bmatrix} 2 \\ 8 \\ 8 \end{bmatrix}$. The coordinates of the same point as seen from the B coordinate system are $\begin{bmatrix} 12 \\ 20 \\ -8 \end{bmatrix}$.

   The B coordinate system can be obtained by initially aligning it with the A system, translating it to a point, and then rotating 40 degrees about the Z axis. Determine the coordinates of the origin of the B coordinate system measured in terms of the A coordinate system.

3. Coordinate system B is initially aligned with coordinate system A. It is then rotated 30 degrees about an axis that is parallel to the X axis but that passes through the point $\begin{bmatrix} 10 & 20 & 10 \end{bmatrix}^T$. Coordinate system C is initially aligned with coordinate system A. It is then rotated 60 degrees about an axis $\begin{bmatrix} 2 & 4 & 6 \end{bmatrix}^T$ that passes through the origin. Determine the transformation matrix that relates the C and B coordinate systems, i.e., $^B_C T$.

4. Coordinate systems A and B are initially aligned. Coordinate system B is then rotated by an angle of 35 degrees about its X axis. It is then rotated 120 degrees about its new Y axis. You wish to return coordinate system B to its original orientation (aligned with coordinate system A) by performing one rotation. About what axis and by what angle should coordinate system B be rotated?

5. The following information is given for a robot manipulator:

$$_0\phi_1 = 70°$$
$$\alpha_1 = 50° \qquad a_1 = 0 \qquad _1S_2 = 50 \text{ cm} \qquad _1\theta_2 = 120°$$
$$\alpha_2 = 90° \qquad a_2 = 20 \text{ cm} \qquad _2S_3 = 35 \text{ cm} \qquad _2\theta_3 = 90°.$$

   Determine the coordinates of the origin of the standard coordinate system attached to link 3 in terms of the standard fixed coordinate system. Determine the coordinates of the vectors $^2S^3$ and a^3 in terms of the fixed coordinate system.

3   Statics of a Rigid Body

A *dyname* determined by its six linear
coordinates represents the effect produced by two
forces not intersecting each other, the points
acted upon not being regarded.

Plücker (1866)

The canonical form to which a system of forces
acting on a rigid body can be reduced is a *wrench*
on a screw.

Ball (1900)

3.1   Introduction

In this chapter, it will be shown that forces can be defined as a force magnitude
multiplied by the coordinates of a line (where the direction vector of the line is a
unit vector). Moments will be defined as the magnitude of the moment multiplied by
the coordinates of a line at infinity (where the direction of the moment term is a unit
vector). The net force and torque acting on a body can be simply calculated by adding
the individual force and moment coordinates. It will be shown that this resultant can
be elegantly represented by a single force acting along some line combined with a
moment whose direction is parallel to the line. This combination is called a *wrench*.
The wrench is a force magnitude multiplied by the coordinates of a *screw*. Some think
of a screw as a *line with a pitch*, where the pitch is the ratio of the magnitude of the
moment to the magnitude of the force. However, it is actually more correct to say *a
screw is a screw*[1] and *a line is a screw with zero pitch*.

3.2   The Coordinates of a Force

Figure 3.1 illustrates a force with magnitude f acting upon a rigid body. The force is
acting on a line $ with ray coordinates $\hat{s} = \{S; S_{OL}\}$, where $|S| = 1$. The force f can
be expressed as a scalar multiple $f S$ of the unit vector S, which is bound to the line $.

---

[1] See Ball (1900), and Hunt (1978).

In order to add forces, it is necessary to introduce some reference point O. The moment of the force f about this reference point, i.e., m_O, can be written as $m_O = r \times f$, where r is a vector to any point on the line $\$$. This moment may also be expressed as a scalar multiple $f S_{OL}$, where S_{OL} is the moment vector of the line $\$$, i.e., $S_{OL} = r \times S$. The action of the force upon the body can, thus, be elegantly expressed as a scalar multiple of the unit line vector, and the coordinates for the force are given by

$$\hat{w} = f\hat{s} = f\{S; S_{OL}\}, \tag{3.1}$$

where $S \cdot S = 1$ and $S \cdot S_{OL} = 0$. Equation (3.1) can be expressed in the form

$$\hat{w} = f\hat{s} = \{f; m_O\}. \tag{3.2}$$

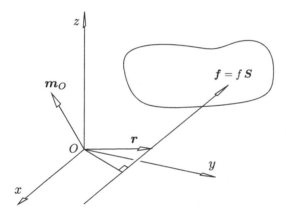

Figure 3.1 Representation of a force

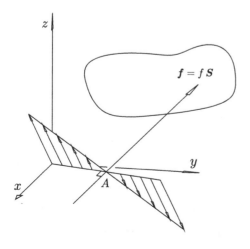

Figure 3.2 Variation in moment vector as reference point moves

Figure 3.2 illustrates the variation of the moment vector m_O as the reference point O changes. When the reference point is coincident with A, $m_O = 0$ and the coordinates of the force are $\{f; 0\}$. Clearly, f is a line bound vector that is invariant with a change of coordinate systems, while m_O is origin dependent.

3.3   The Coordinates of a Couple

Consider the resultant of a pair of equal and opposite forces that act along different lines of action upon a rigid body. These forces are coplanar with coordinates $\{f; m_{O1}\} = f\{S_1; S_{OL1}\}$ and $\{-f; m_{O2}\} = f\{-S_1; S_{OL2}\}$, where $|S_1| = 1$. Figure 3.3 shows the pair of lines whereby the plane formed by the pair of lines passes through the origin. This is done without loss of generality for ease of visualization. The vectors from the reference point O perpendicular to each of the lines of action are given by (1.59) as

$$p_1 = S_1 \times S_{OL1}, \tag{3.3}$$

$$p_2 = -S_1 \times S_{OL2}. \tag{3.4}$$

The moment exerted by the pair of forces, i.e., m, is calculated as

$$m = (p_1 - p_2) \times f S_1. \tag{3.5}$$

Substituting (3.3) and (3.4) into (3.5) and rearranging yields

$$\begin{aligned} m &= f(S_1 \times S_{OL1}) \times S_1 + f(S_1 \times S_{OL2}) \times S_1 \\ &= f S_1 \times (S_{OL1} \times S_1) + f S_1 \times (S_{OL2} \times S_1), \end{aligned} \tag{3.6}$$

and expanding the vector triple products[2] yields

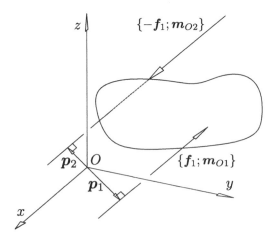

Figure 3.3 Pair of equal magnitude and opposite direction forces

[2] Note that $\mathbf{a} \times (\mathbf{b} \times \mathbf{c}) = (\mathbf{a} \cdot \mathbf{c})\mathbf{b} - (\mathbf{a} \cdot \mathbf{b})\mathbf{c}$.

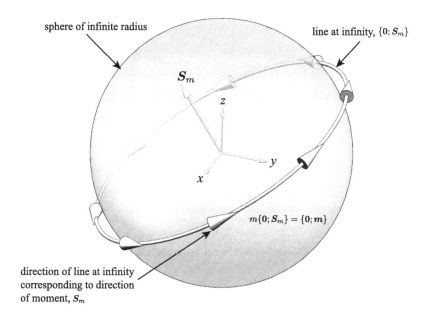

sphere of infinite radius

line at infinity, $\{0; \boldsymbol{S}_m\}$

$\boldsymbol{S}_m$

z

y

x

$m\{0; \boldsymbol{S}_m\} = \{0; \boldsymbol{m}\}$

direction of line at infinity
corresponding to direction
of moment, $\boldsymbol{S}_m$

Figure 3.4 Conceptualization of a moment as a line at infinity multiplied by a moment magnitude

$$\boldsymbol{m} = f(\boldsymbol{S}_1 \cdot \boldsymbol{S}_1)\boldsymbol{S}_{OL1} + f(\boldsymbol{S}_1 \cdot \boldsymbol{S}_1)\boldsymbol{S}_{OL2}$$
$$= f(\boldsymbol{S}_{OL1} + \boldsymbol{S}_{OL2})$$
$$= \boldsymbol{m}_{O1} + \boldsymbol{m}_{O2}. \tag{3.7}$$

The moment, or couple, caused by the pair of forces can be considered as equivalent to a line at infinity with coordinates $\{0; \boldsymbol{S}_m\}$, $|\boldsymbol{S}_m| = 1$, where $\boldsymbol{S}_m$ is dimensionless, times a moment magnitude, m. The magnitude m must satisfy the relation $m\boldsymbol{S}_m = \boldsymbol{m}$. The moment $\{0; \boldsymbol{m}\} = m\{0; \boldsymbol{S}_m\}$ is depicted in Figure 3.4. Thus, a pure couple can be expressed as a scalar multiple of a line at infinity. It is not a line bound vector nor are its coordinates dependent on the selection of the reference point O.

3.4   Translation of a Force: Equivalent Force/Couple Combination

Figure 3.5(a) shows a rigid body upon which is acting a force $\{\boldsymbol{f}_1; \boldsymbol{m}_{O1}\}$ whose coordinates are expressed in terms of the reference point O. Equal and opposite collinear forces whose coordinates are $\{\boldsymbol{f}_1; \boldsymbol{m}_{O2}\}$ and $-\{\boldsymbol{f}_1; \boldsymbol{m}_{O2}\}$ are applied to the body, as shown in Figure 3.5(b), and it is apparent that the net resultant force and couple acting on the body are the same as in Figure 3.5(a). The forces $\{\boldsymbol{f}_1; \boldsymbol{m}_{O1}\}$ and $-\{\boldsymbol{f}_1; \boldsymbol{m}_{O2}\}$ may be replaced by a couple that is equal to the sum of the moments of the two line bound forces about point O as per equation (3.7), i.e., for this case, $\boldsymbol{m} = \boldsymbol{m}_{O1} - \boldsymbol{m}_{O2}$. The combination of the force $\{\boldsymbol{f}_1; \boldsymbol{m}_{O2}\}$ and the couple $\{0; \boldsymbol{m}\}$,

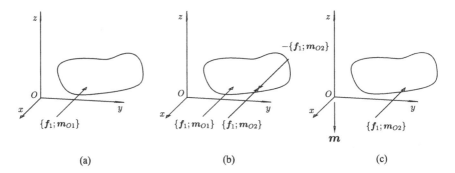

Figure 3.5 Translation of a force; $\{f_1; m_{O1}\} = \{f_1; m_{O2}\} + \{0; m\}$: (a) original force, (b) transitioning to a new parallel force, and (c) the translated force/couple result

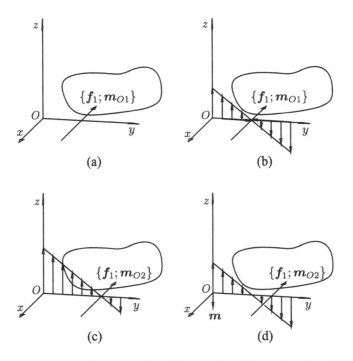

Figure 3.6 Translation of a force; $\{f_1; m_{O1}\} = \{f_1; m_{O2}\} + \{0; m\}$: (a) original force, (b) original force showing moment distribution, (c) translated force with its moment distribution, and (d) translated force/couple shown with original moment distribution

as shown in Figure 3.5(c), has the same effect on the rigid body as the original force $\{f_1; m_{O1}\}$.

This same problem may be restated in a different manner. Shown in Figure 3.6(a) is a force $\{f_1; m_{O1}\}$, which is acting on some rigid body. The moment of the force about point O is m_{O1}. The moment of the force about points other than O can be readily determined. Figure 3.6(b) shows a linear variation in the moment vector field. In Figure 3.6(c), the force has been translated to a new line of action, and its coordinates may be written as $\{f_1; m_{O2}\}$. The moment of the force about point O is now m_{O2}. Figure 3.6(c) also shows the linear variation of the moment vector field caused by the force acting along its new line of action. Figure 3.6(d) shows the sum of this vector field with the moment $m = m_{O1} - m_{O2}$, which yields the exact same moment vector field as shown in Figure 3.6(b). The overall moment vector field caused by the force about its original line of action is identical to the resultant moment vector field caused by the force acting about its new line of action summed with $m_{O1} - m_{O2}$. Because of this, the effect on the rigid body of the force acting along its original line of action is identical to the effect on the rigid body of the force acting on its new line of action together with the moment $m = m_{O1} - m_{O2}$.

Thus, a line bound force is equivalent to the combination of a parallel force of equal magnitude and a new couple. The magnitude of the new couple is equal to the magnitude of the force times the perpendicular distance between the force's original line of action and its new line of action. The direction of the couple is perpendicular to the plane formed by the force's original line of action and its new line of action.

3.5 A Dyname and a Wrench

Figure 3.7(a) illustrates an arbitrary system of forces with coordinates $\{f_1; m_{O1}\}$, $\{f_2; m_{O2}\}, \ldots, \{f_n; m_{On}\}$ acting upon a rigid body. It is assumed at the outset that a reference point O has been chosen, the magnitudes of the forces $f_1, f_2, \ldots, f_n$ are specified, and the unit coordinates of the lines of action of the forces $\{S_1; S_{OL1}\}$, $\{S_2; S_{OL2}\}, \ldots, \{S_n; S_{OLn}\}$ are known. In Figure 3.7(b), the forces have been translated to point O, and moments $m_{O1}, m_{O2}, \ldots, m_{On}$ have been introduced to yield an equivalent system of forces and torques that act on the rigid body.

The net force acting on the rigid body is given by

$$f = \sum_{i=1}^{n} f_i, \tag{3.8}$$

and the line of action of this force passes through point O. In addition to this force there is a moment acting on the rigid body, which is given by

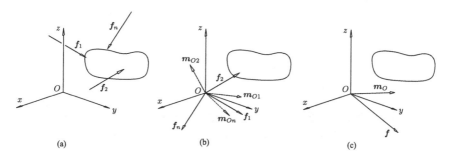

(a)   (b)   (c)

Figure 3.7 Dyname $\{f; m_O\}$: (a) a number of forces acting, (b) forces moved to origin with associated couples, and (c) equivalent dyname

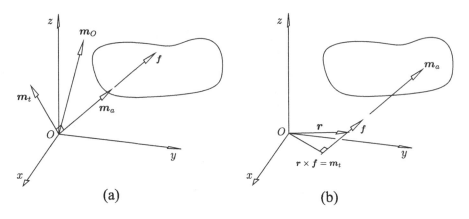

Figure 3.8 (a) Dyname $\{f;0\} + \{0;m_O\}$. (b) Wrench $\{f;m_t\} + \{0;m_a\}$

$$m_O = \sum_{i=1}^{n} m_{Oi}. \qquad (3.9)$$

The coordinates of the moment may be written as $\{0;m_O\}$ as per Section 3.3, where the moment can be considered as an infinitesimal force acting at infinity whose moment vector relative to the reference point O is m_O.

The net force f (which passes through point O and whose coordinates are written as $\{f;0\}$) and moment m_O (which is a non-line-bound vector whose coordinates are written as $\{0;m_O\}$) are shown in Figure 3.7(c), and the combination of this force and moment is equivalent to the original set of forces. The coordinates of the resultant of the force and moment may be written as the sum of $\{f;0\}$ and $\{0;m_O\}$ as

$$\hat{w} = \{f;m_O\}. \qquad (3.10)$$

In general, f and m_O will not be perpendicular and the quantity with coordinates $\hat{w} = \{f;m_O\}$, $f \cdot m_O \neq 0$, was defined as a dyname by Plücker.[3]

Since in general $f \cdot m_O \neq 0$, it is not possible to translate the line of action of f through some point other than point O and have the translated force produce the same net effect on the rigid body as the original dyname. The moment m_O, however, may be resolved into two components m_a and m_t as

$$m_O = m_a + m_t, \qquad (3.11)$$

which are respectively parallel to f and perpendicular to f (see Figure 3.8(a)). The moment m_a may be determined as

$$m_a = (m_O \cdot S)\,S, \qquad (3.12)$$

where S is a unit vector in the direction of the resultant force f. The moment m_t is then determined as

---

[3] See Plücker (1866).

$$m_t = m_O - m_a. \tag{3.13}$$

The line of action of force f may now be translated so that the force with coordinates $\{f; m_t\}$ plus the moment $\{0; m_a\}$ (see Figure 3.8(b)) is equivalent to the dyname $\{f; 0\} + \{0; m_O\}$. The dyname can, thus, be represented uniquely by a force f acting on the line of action $\{S; S_{Ot}\}$, where $S_{Ot} = \frac{m_t}{f}$ and a parallel couple m_a. *This representation is called a wrench and is due to Ball.*

The wrench $\hat{w}$, which is equivalent to the dyname $\{f; m_O\}$, may be written as

$$\hat{w} = \{f; m_O\} = \{f; m_t\} + \{0; m_a\}. \tag{3.14}$$

It is preferable to express the right side of (3.14) in terms of f and m_O. Firstly, since m_a is parallel to f, we may write

$$m_a = h f, \tag{3.15}$$

where h is a non-zero scalar which is called the pitch of the wrench. The pitch h clearly is the ratio of the magnitude of the moment m_a divided by the magnitude of the force f and can be calculated by performing a scalar product of both sides of (3.15) with f and then solving for h as

$$h = \frac{f \cdot m_a}{f \cdot f}. \tag{3.16}$$

From (3.11)

$$f \cdot m_O = f \cdot (m_a + m_t) = f \cdot m_a \tag{3.17}$$

and from (3.16) and (3.17)

$$h = \frac{f \cdot m_O}{f \cdot f}. \tag{3.18}$$

Clearly the pitch h is an invariant quantity, and it has units of length. Substituting (3.13) into (3.14) allows the wrench $\hat{w}$ to be written as

$$\hat{w} = \{f; m_O - m_a\} + \{0; m_a\} \tag{3.19}$$

and then substituting (3.15) gives

$$\hat{w} = \{f; m_O - h f\} + \{0; h f\}. \tag{3.20}$$

The homogeneous coordinates for the line of action of the wrench are $\frac{1}{f}\{f; m_O - h f\}$, and the equation for the line is, therefore,

$$r \times \frac{1}{f} f = \frac{1}{f} (m_O - h f). \tag{3.21}$$

In the same way as the action of a force upon a body can be expressed as a scalar multiple of a unit line vector, a wrench acting on a body can be elegantly expressed as a scalar multiple of a unit screw $f\hat{s}$, where

$$\hat{s} = \{S; S_O\} \tag{3.22}$$

and where $S \cdot S = 1$. The pitch of the screw is given by

$$h = S \cdot S_O. \tag{3.23}$$

From (3.22),

$$\hat{s} = \{S; S_O\} = \{S; S_O - hS\} + \{0; hS\}. \tag{3.24}$$

The Plücker coordinates for the screw axis are $\{S; S_{OL}\} = \{S; S_O - hS\}$, and the equation of the axis is

$$r \times S = S_O - hS. \tag{3.25}$$

A screw can, therefore, be regarded as a line with a pitch.

3.5.1   Sample Problem

Three forces and one moment are acting on a rigid body. The directions, magnitudes, and a point on the line of action of the forces are given as

$$
\begin{aligned}
S_{dir1} &= [1, 2, 3]^T & f_1 &= 10\text{ N} & r_1 &= [0.5, 0.25, 1]^T \text{ m} \\
S_{dir2} &= [-2, 1, 2]^T & f_2 &= 15\text{ N} & r_2 &= [2.5, -1, 1.5]^T \text{ m} \\
S_{dir3} &= [1, -1, -1]^T & f_3 &= 18\text{ N} & r_3 &= [-3.5, 0, -3]^T \text{ m.}
\end{aligned}
$$

The direction and magnitude of the moment are given as

$$S_{dir4} = [3, 0, 2]^T \qquad m_4 = 36 \; Nm.$$

(i)   *Determine the resulting dyname $\{f; m_O\}$ that is acting on the body.*

The coordinates of the three forces may be determined from $f_i \{S_i; r_i \times S_i\}$, $i = 1 \ldots 3$, where S_i is now written as a unit vector as

$$\{f_1; m_{O1}\} = \{2.6726, 5.3452, 8.0178; -3.3408, -1.3363, 2.0045\},$$

$$\{f_2; m_{O2}\} = \{-10, 5, 10; -17.5, -40, 2.5\}.$$

$$\{f_3; m_{O3}\} = \{10.3923, -10.3923, -10.3923; -31.1769, -67.5500, 36.3731\},$$

where the units of the first three components are N, and the units of the last three components are Nm. The coordinates of the moment may be written as

$$\{0; m_4\} = \{0, 0, 0; 29.9538, 0, 19.9692\},$$

where, as for the force coordinates, the units of the first three components are N, and the units of the last three components are Nm.

The net resulting dyname may be determined as

$$
\begin{aligned}
\{f; m_O\} &= \{f_1; m_{O1}\} + \{f_2; m_{O2}\} + \{f_3; m_{O3}\} + \{0; m_4\} \\
&= \{3.0649, -0.0471, 7.6255; -22.0639, -108.8863, 60.8467\},
\end{aligned}
$$

where again the units of the first three components are N, and the units of the last three components are Nm. This dyname can be interpreted as a force passing through the origin of magnitude 8.2186 N in the direction $[0.3729, -0.0057, 0.9278]^T$

together with a moment of magnitude 126.6703 Nm in the direction $[-0.1742, -0.8596, 0.4804]^T$.

(ii) *Determine the pitch, magnitude, and line of action of the equivalent wrench.*
The pitch of the equivalent wrench can be determined from (3.18) as

$$h = \frac{\boldsymbol{f} \cdot \boldsymbol{m}_O}{\boldsymbol{f} \cdot \boldsymbol{f}}$$
$$= \frac{[3.0649, -0.0471, 7.6255] \cdot [-22.0639, -108.8863, 60.8467]}{[3.0649, -0.0471, 7.6255] \cdot [3.0649, -0.0471, 7.6255]}$$
$$= 5.9441 \text{ m}.$$

The magnitude of the equivalent wrench is determined as

$$f = |\boldsymbol{f}| = 8.2186 \text{ N}.$$

The coordinates of the screw $\{S; S_O\}$ are obtained as

$$\{S; S_O\} = \frac{1}{f}\{\boldsymbol{f}; \boldsymbol{m}_O\},$$

and the coordinates of the line of action are determined from (3.24) as

$$\{S; S_{OL}\} = \{S; S_O - hS\}$$
$$= \{0.3729, -0.0057, 0.9278; -4.9013, -13.2148, 1.8884\},$$

where first three coordinates are dimensionless, and the last three coordinates have units of m.

3.6   Transformation of Screw Coordinates

Figure 3.9 shows a screw $\$$ whose coordinates are assumed to be known in coordinate system B as $^B\$ = \{^BS; {}^BS_O\} = \{^BS; {}^B\boldsymbol{r}_1 \times {}^BS + h \, {}^BS\}$. It is desired to determine the coordinates of this screw in terms of a coordinate system A where the relationship between the coordinate systems is known as ^A_BT, which is the 4×4 transformation matrix defined by ^A_BR and $^A\boldsymbol{P}_{Bo}$, the 3×3 orientation matrix, and the coordinates of the origin of the B coordinate system measured with respect to the A system.

The Plücker coordinates of the line along the screw may be written in terms of the B coordinate system as

$$\{^BS; {}^BS_{OL}\} = \{^BS; {}^BS_O - h \, {}^BS\}, \tag{3.26}$$

where

$$h = \frac{^BS \cdot {}^BS_O}{^BS \cdot {}^BS}. \tag{3.27}$$

From (2.65), the coordinates of this line may be expressed in terms of the A coordinate system as

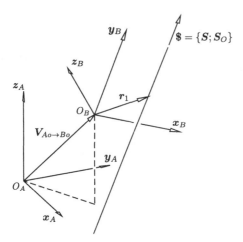

Figure 3.9 Transformation of screw coordinates

$$\{^{A}S;\ ^{A}S_{OL}\} = \{^{A}_{B}R\ ^{B}S;\ ^{A}P_{Bo} \times ^{A}_{B}R\ ^{B}S + ^{A}_{B}R\ ^{B}S_{OL}\}. \tag{3.28}$$

The pitch of a screw is an invariant quantity with respect to a change of origin. In other words, the pitch of a screw remains the same, no matter from what coordinate system it is observed. Thus, the coordinates of the screw $ may be expressed in terms of the A coordinate system as

$$\{^{A}S;\ ^{A}S_{O}\} = \{^{A}S;\ ^{A}S_{OL}\} + \{0; h\ ^{A}S\} \tag{3.29}$$

$$= \{^{A}S;\ ^{A}S_{OL}\} + \{0; h\ ^{A}_{B}R\ ^{B}S\}$$

$$= \{^{A}_{B}R\ ^{B}S;\ ^{A}P_{Bo} \times ^{A}_{B}R\ ^{B}S + ^{A}_{B}R\ ^{B}S_{OL} + h\ ^{A}_{B}R\ ^{B}S\}.$$

Recognizing that $^{B}S_{OL} + h\ ^{B}S = ^{B}S_{O}$ yields

$$\{^{A}S;\ ^{A}S_{O}\} = \{^{A}_{B}R\ ^{B}S;\ ^{A}_{B}R\ ^{B}S_{O} + ^{A}P_{Bo} \times ^{A}_{B}R\ ^{B}S\}. \tag{3.30}$$

3.6.1   Sample Problem

The coordinates of two wrenches are given as

$$\hat{w}_1 = \{20, 10, -6; 120, 30, -30\},$$

$$\hat{w}_2 = \{10, -4, 8; -100, 100, -140\},$$

where the units of the first three components are N, and the last three components are Nm.

Determine the following:

(a) force magnitude of each wrench, f_1 and f_2

(b) the coordinates of the two screws, where the direction of the screw is a unit vector

(c) the pitch of each screw, h_1 and h_2

(d) the Plücker line coordinates of the lines of action of the two wrenches, $\{S_1; S_{OL1}\}$ and $\{S_2; S_{OL2}\}$

(e) the angle α_{12} between the lines of action of the two wrenches as measured in a right-hand sense from the direction of the first wrench to the second wrench about the vector $\boldsymbol{a}_{12}$, which is defined as the unit vector parallel to $S_1 \times S_2$

(f) the perpendicular distance a_{12} between the lines of action of the two wrenches

(g) the coordinates of the point $\boldsymbol{r}_{E1}$, which is defined as the intersection point of the line of action of the first wrench with the line perpendicular to the lines of action of the two wrenches

(h) the coordinates of the two wrenches expressed in a coordinate system whose origin is at point $\boldsymbol{r}_{E1}$, whose Z axis is along the direction of the perpendicular line and whose X axis is along the direction of the line of action of the first wrench.

Solution:

(a) *Determine the force magnitude of each wrench, f_1 and f_2.*

The force magnitude of the two screws can be calculated as

$$f_1 = \sqrt{20^2 + 10^2 + (-6)^2} = 23.15 \text{ N},$$
$$f_2 = \sqrt{10^2 + (-4)^2 + 8^2} = 13.42 \text{ N}.$$

(b) *Determine the coordinates of the two screws where the direction of the screw is a unit vector.*

The coordinates of the two screws can be determined by dividing the wrench coordinates by the force magnitudes to yield

$$\$_1 = \{S_1; S_{O1}\} = \{0.8639, 0.4319, -0.2592; 5.1832, 1.2958, -1.2958\},$$
$$\$_2 = \{S_2; S_{O2}\} = \{0.7454, -0.2981, 0.5963; -7.4536, 7.4536, -10.4350\},$$

where the first three components are dimensionless, and the second three have units of m.

(c) *Determine the pitch of each screw, h_1 and h_2.*

Since $|S_1| = |S_2| = 1$, the pitch of each screw may be calculated as

$$h_i = S_i \cdot S_{Oi}, i = 1 \dots 2.$$

Thus,

$$h_1 = 5.3731 \text{ m},$$
$$h_2 = -14.0000 \text{ m}.$$

(d) *Determine the Plücker line coordinates of the lines of action of the two wrenches, $\{S_1; S_{OL1}\}$ and $\{S_2; S_{OL2}\}$.*

The Plücker line coordinates of the lines of action of the two screws can be obtained as

$$\{S_i; S_{OLi}\} = \{S_i; S_{Oi} - h_i S_i\}.$$

Thus,

$$\{S_1; S_{OL1}\} = \{0.8639, 0.4319, -0.2592; 0.5415, -1.0254, 0.0967\},$$
$$\{S_2; S_{OL2}\} = \{0.7454, -0.2981, 0.5963; 2.9814, 3.2796, -2.0870\},$$

where the first three components are dimensionless, and the last three components have units of m.

(e) *Determine the angle α_{12} between the lines of action of the two wrenches as measured in a right-hand sense from the direction of the first wrench to the second wrench about the vector a_{12}, which is defined as the unit vector parallel to $S_1 \times S_2$.*

The cosine of α_{12} can be obtained from

$$\cos \alpha_{12} = S_1 \cdot S_2 = 0.3606.$$

The unit vector a_{12}, which is perpendicular to S_1 and S_2, is determined as

$$a_{12} = \frac{S_1 \times S_2}{|S_1 \times S_2|} = 0.1933\, i - 0.7594\, j - 0.6212\, k,$$

and the sine of the angle α_{12} is determined from (1.139) as

$$\sin \alpha_{12} = S_1 \times S_2 \cdot a_{12} = 0.9327.$$

The angle α_{12} is determined as

$$\alpha_{12} = \text{atan2}(\sin \alpha_{12}, \cos \alpha_{12}) = 1.2019\, rad = 68.86°.$$

(f) *Determine the perpendicular distance a_{12} between the lines of action of the two wrenches.*

The mutual moment of the two lines can be written as per equation (1.141) as

$$S_1 \cdot S_{OL2} + S_2 \cdot S_{OL1} = -a_{12} \sin \alpha_{12}.$$

The left side of this equation evaluates to

$$S_1 \cdot S_{OL2} + S_2 \cdot S_{OL1} = 5.300 \text{ m.}$$

The perpendicular distance between the lines of action of the screws is now obtained from the mutual moment equation as

$$a_{12} = -\frac{S_1 \cdot S_{OL2} + S_2 \cdot S_{OL1}}{\sin \alpha_{12}} = -5.6821 \text{ m.}$$

The negative value indicates that to move from the first line to the second, one would move a distance of 5.6821 meters in the direction opposite to a_{12}.

(g) *Determine the coordinates of the point r_{E1}, which is defined as the intersection point of the line of action of the first wrench with the line perpendicular to the lines of action of the two wrenches.*

The point r_{E1} can be determined from equation (1.160) as

$$r_{E1} = \frac{S_{OL1} \times (S_{OL2} - a_{12} a_{12} \times S_2)}{S_{OL1} \cdot S_2}$$
$$= -1.0706\, i - 0.6472\, j - 0.8654\, k,$$

where the components of this vector have units of meters.

(h) *Determine the coordinates of the two wrenches expressed in a coordinate system whose origin is at point r_{E1}, whose Z axis is along the direction of the perpendicular line, and whose X axis is along the direction of the line of action of the first wrench.*

The reference coordinate system will be referred to as the A coordinate system. The coordinates of the lines of action of the two wrenches were evaluated in this reference system as

$$^A\$_1 = \{^A S_1; \,^A S_{O1}) = \{0.8639, 0.4319, -0.2592; 5.1832, 1.2958, -1.2958\},$$
$$^A\$_2 = \{^A S_2; \,^A S_{O2}) = \{0.7454, -0.298, 0.5963; -7.4536, 7.4536, -10.435\},$$

where the first three components are dimensionless, and the last three components have units of meters. The B coordinate system will be initially aligned with the A coordinate system and is then translated to the point r_{E1}. Thus, the coordinates of the origin point of the B coordinate system as evaluated by an observer in the A coordinate system is

$$^A P_{Bo} = \,^A r_{E1}.$$

After the translation, its coordinate axes will then be oriented so that the Z axis is along a_{12} and the X axis is along S_1. The orientation matrix $^A_B R$, which relates the orientation of the two coordinate systems, may be written as

$$^A_B R = \begin{bmatrix} ^A S_1 & ^A a_{12} \times ^A S_1 & ^A a_{12} \end{bmatrix}$$

$$= \begin{bmatrix} 0.8639 & 0.4652 & 0.1933 \\ 0.4319 & -0.4866 & -0.7594 \\ -0.2592 & 0.7395 & -0.6212 \end{bmatrix}.$$

By exchanging the subscripts in (3.30), the coordinates of a screw known in the A coordinate system may be written in terms of the B coordinate system as

$$\{^B S; \,^B S_O\} = \{^B_A R \,^A S; \,^B_A R \,^A S_O + \,^B P_{Ao} \times \,^B_A R \,^A S\}.$$

Equation (2.20) gives $^B_A R = \,^A_B R^T$, and (2.22) gives $^B P_{Ao} = -\,^A_B R^T \,^A P_{Bo}$. Thus,

$$\{^B S; \,^B S_O\} = \{^A_B R^T \,^A S; \,^A_B R^T \,^A S_O - \,^A_B R^T \,(^A P_{Bo} \times \,^A S)\}.$$

Applying this equation to the two screws whose coordinates are known in the A system gives

$$\{^B S_1; \,^B S_{O1}\} = \{1, 0, 0; 5.3731, 0, 0\}$$
$$\{^B S_2; \,^B S_{O2}\} = \{0.3606, 0.9328, 0; 0.2518, -15.1070, 0\},$$

where the first three components are dimensionless, and the last three components have units of meters. Lastly, the coordinates of the two wrenches may be determined in the B coordinate system by multiplying each of the screw coordinates $\{^B S_1; \,^B S_{O1}\}$ and $\{^B S_2; \,^B S_{O2}\}$ by the force magnitudes f_1 and f_2, respectively, to give

$$^B\hat{w}_1 = f_1\{^BS_1; \ ^BS_{O1}\} = \{23.15, 0, 0; 124.40, 0, 0\}$$
$$^B\hat{w}_2 = f_2\{^BS_2; \ ^BS_{O2}\} = \{4.84, 12.51, 0; 3.38; -202.68, 0\},$$

where the first three components of each wrench have units of N, and the last three components have units of Nm.

3.7   Forward and Reverse Static Analysis of In-Parallel Platform Devices

The term 'in-parallel' was coined by Hunt (1983) to define parallel mechanisms with a fixed base connected to a moving platform by six legs, each of which is the same kinematic chain. Further, the same joint in each chain is actuated. Figure 3.10 illustrates a 6-6 in-parallel device, so-called because there are six connecting points in the base and six connecting points in the moving platform. Each leg is an S-P-U serial kinematic chain, where the letter S is used to represent a spherical (ball and socket) joint, the letter U is used to represent a Hooke joint, and the letter P represents a prismatic joint which, in each leg, is actuated. There are $3 + 1 + 2 = 6$ freedoms in each leg.

The forward static analysis consists of computing the resultant wrench $\hat{w} = \{f; m_O\}$ due to the six forces generated in the legs acting upon the moving platform. Now $\hat{w}$ can be expressed in the form

$$\hat{w} = \{f_1; m_{O1}\} + \{f_2; m_{O2}\} + \cdots + \{f_6; m_{O6}\} \tag{3.31}$$

or

$$\hat{w} = f_1\{S_1; S_{OL1}\} + f_2\{S_2; S_{OL2}\} + \cdots + f_6\{S_6; S_{OL6}\}, \tag{3.32}$$

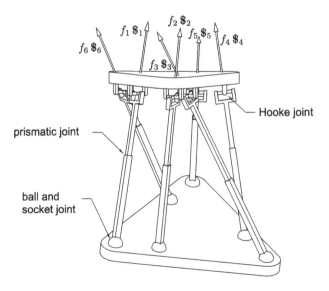

Figure 3.10 Six forces acting on a body via six in-parallel SPS serial chains

where $\{S_i; S_{OLi}\}$, $i = 1\ldots 6$ are the Plücker coordinates of the six legs. It is convenient to express (3.32) in the form

$$\hat{w} = f_1 \begin{bmatrix} S_1 \\ S_{OL1} \end{bmatrix} + f_2 \begin{bmatrix} S_2 \\ S_{OL2} \end{bmatrix} + \cdots + f_6 \begin{bmatrix} S_6 \\ S_{OL6} \end{bmatrix}, \tag{3.33}$$

which may be written as

$$\hat{w} = j\,\lambda, \tag{3.34}$$

where j is a 6×6 matrix given as

$$j = \begin{bmatrix} S_1 & S_2 & S_3 & S_4 & S_5 & S_6 \\ S_{OL1} & S_{OL2} & S_{OL3} & S_{OL4} & S_{OL5} & S_{OL6} \end{bmatrix}, \tag{3.35}$$

and λ is a column vector given as

$$\lambda = \begin{bmatrix} f_1 \\ f_2 \\ f_3 \\ f_4 \\ f_5 \\ f_6 \end{bmatrix}. \tag{3.36}$$

The geometry of the platform, i.e., the position and orientation of the top platform relative to the base, is assumed to be known at the outset so that the six columns of j are known. Further, the magnitude f_i, $i = 1\ldots 6$, of each of the leg forces is known. The coordinates of the resultant wrench can be computed from (3.34). This resultant wrench may be written as

$$\hat{w} = \begin{bmatrix} f \\ m_O \end{bmatrix} = \begin{bmatrix} L \\ M \\ N \\ P \\ Q \\ R \end{bmatrix}. \tag{3.37}$$

The magnitude of the resultant wrench is given by

$$f = |f| = \sqrt{L^2 + M^2 + N^2}. \tag{3.38}$$

The pitch h is determined from (3.18) as

$$h = \frac{f \cdot m_O}{f \cdot f} = \frac{LP + MQ + NR}{L^2 + M^2 + N^2}. \tag{3.39}$$

Finally, the coordinates for the line of action of the resultant wrench are given by

$$S = \begin{bmatrix} \dfrac{L}{|f|}, \dfrac{M}{|f|}, \dfrac{N}{|f|} \end{bmatrix}^T, \tag{3.40}$$

$$S_{OL} = \begin{bmatrix} \dfrac{P - hL}{|f|}, \dfrac{Q - hM}{|f|}, \dfrac{R - hN}{|f|} \end{bmatrix}. \tag{3.41}$$

This completes the forward static analysis. Clearly, for equilibrium, an external wrench with an equal and opposite magnitude $|f|$ must be applied to the platform.

Conversely, when the position and orientation of the top platform is known relative to the base and an external wrench or force is applied to the top platform, it is required to determine the magnitudes $f_1, f_2, \ldots, f_6$ of the connector forces. *This is called the reverse or inverse static analysis.* This is accomplished by solving (3.34) for λ as

$$\lambda = j^{-1}\,\hat{w}, \tag{3.42}$$

where j^{-1} is the inverse of j. This computation cannot be performed when the rank of j is less than six, for which the Plücker coordinates of the lines along the six leg connectors become linearly dependent and the platform is in a singularity position.

3.8   Forward and Reverse Static Analysis of a Serial Manipulator

Figure 3.11 shows an external wrench, $\hat{w}_{ext}$, applied to the end effector of a serial manipulator as well as actuator efforts at each of the joints. For a revolute joint, the actuator effort will be the magnitude of the applied torque. For a prismatic joint, the actuator effort will be the magnitude of the applied force. The forward static analysis consists of computing the resultant wrench $\hat{w}_{ext} = \{f_{ext}; m_{ext}\}$ that is acting on the manipulator end effector due to the torques and/or forces acting at the joints. In the reverse static analysis, the external wrench, $\hat{w}_{ext}$, is given and the joint actuation

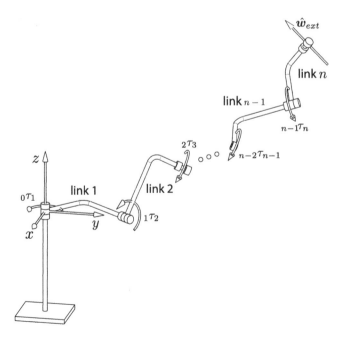

Figure 3.11 External wrench applied to end effector of a serial manipulator

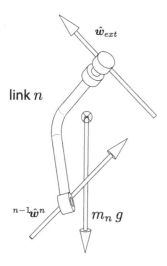

link n

$\hat{w}_{ext}$

$^{n-1}\hat{w}^n$

$m_n \, g$

Figure 3.12 Free body diagram of end effector, link n

efforts are to be calculated that hold the system in static equilibrium. In both cases, it is assumed that the constant mechanism parameters and the values for the variable joint parameters are known.

The derivation begins by considering the free body diagram of the last link of the manipulator (see Figure 3.12). The external wrench $\hat{w}_{ext}$ is shown together with the gravity force, $m_n \, g$, acting through the link's center of mass point. The wrench applied due to the reaction forces and torques and the joint actuation effort are labeled as $^{n-1}\hat{w}^n$. Since the link is in static equilibrium,

$$^{n-1}\hat{w}^n = -\hat{w}_{ext} - m_n \, g \, \$_{nG}, \tag{3.43}$$

where $\$_{nG}$ represents the line in the direction of gravity which passes through the center of mass of link n. This may be written as

$$\$_{nG} = \left[\begin{array}{c} S_{grav} \\ r_{nCM} \times S_{grav} \end{array} \right], \tag{3.44}$$

where S_{grav} is a unit vector in the direction of gravity, r_{nCM} is the coordinates of the center of mass point for link n, and m_n is the mass of link n.

The actuation effort at the joint connecting bodies n and $n - 1$ can be determined from the term $^{n-1}\hat{w}^n = \{^{n-1}f^n; \, ^{n-1}m_O^n\}$. For a prismatic joint, the actuation force can be determined by projecting the force vector onto the direction of the translation, i.e., the unit vector $^{n-1}S^n$, as

$$_{n-1}f_n = {}^{n-1}S^n \cdot {}^{n-1}f^n. \tag{3.45}$$

For a revolute joint, the actuation torque can be determined by projecting the moment associated with $^{n-1}\hat{w}^n$ on the direction of the axis of rotation. Treating $^{n-1}\hat{w}^n$ as a dyname, it can be thought of as a force $^{n-1}f^n$ through the origin point combined with a moment $^{n-1}m^n$. The moment of the force with respect to a point on the joint's axis of rotation may be written as $-r_n \times {}^{n-1}f^n$, where r_n is the vector to any point on the axis of rotation associated with the joint. Projecting the total moment associated with $^{n-1}\hat{w}^n$ onto the direction of the axis of rotation, i.e., the unit vector $^{n-1}S^n$, yields

$$_{n-1}\tau_n = {}^{n-1}S^n \cdot \left({}^{n-1}m^n - r_n \times {}^{n-1}f^n \right). \tag{3.46}$$

For link $n - 1$, the static equilibrium equation may be written as

$$^{n-2}\hat{w}^{n-1} + m_{n-1}\, g\, \$_{(n-1)G} - {}^{n-1}\hat{w}^n = \mathbf{0}. \tag{3.47}$$

A recursive expression can be written for the remaining links as

$$^{i-2}\hat{w}^{i-1} = {}^{i-1}\hat{w}^i - m_{i-1}\, g\, \$_{(i-1)G}, \tag{3.48}$$

where i varies from n to 2.

For a six axis manipulator, six equilibrium equations will exist; one for each of the moving bodies. These equations will be equation (3.43) and five equations associated with (3.48). These six equations are written here as

$$\begin{aligned}
{}^5\hat{w}^6 &= -\hat{w}_{ext} - m_6\, g\, \$_{6G}, \\
{}^4\hat{w}^5 &= {}^5\hat{w}^6 - m_5\, g\, \$_{5G}, \\
{}^3\hat{w}^4 &= {}^4\hat{w}^5 - m_4\, g\, \$_{4G}, \\
{}^2\hat{w}^3 &= {}^3\hat{w}^4 - m_3\, g\, \$_{3G}, \\
{}^1\hat{w}^2 &= {}^2\hat{w}^3 - m_2\, g\, \$_{2G}, \\
{}^0\hat{w}^1 &= {}^1\hat{w}^2 - m_1\, g\, \$_{1G}.
\end{aligned} \tag{3.49}$$

These equations may be written as

$$\begin{aligned}
{}^5\hat{w}^6 &= -\hat{w}_{ext} - g\, m_6\$_{6G}, \\
{}^4\hat{w}^5 &= -\hat{w}_{ext} - g\, (m_6\$_{6G} + m_5\$_{5G}), \\
{}^3\hat{w}^4 &= -\hat{w}_{ext} - g\, (m_6\$_{6G} + m_5\$_{5G} + m_4\$_{4G}), \\
{}^2\hat{w}^3 &= -\hat{w}_{ext} - g\, (m_6\$_{6G} + m_5\$_{5G} + m_4\$_{4G} + m_3\$_{3G}), \\
{}^1\hat{w}^2 &= -\hat{w}_{ext} - g\, (m_6\$_{6G} + m_5\$_{5G} + m_4\$_{4G} + m_3\$_{3G} \\
&\quad + m_2\$_{2G}) \\
{}^0\hat{w}^1 &= -\hat{w}_{ext} - g\, (m_6\$_{6G} + m_5\$_{5G} + m_4\$_{4G} + m_3\$_{3G} \\
&\quad + m_2\$_{2G} + m_1\$_{1G}).
\end{aligned} \tag{3.50}$$

For the reverse static analysis, the external wrench is given, and all six wrenches associated with the joints, i.e., $^{i-1}\hat{w}^i$, can be calculated. The actuation force or torque at each joint can then be calculated from either (3.45) or (3.46) depending on whether the joint is a prismatic or revolute joint. For the forward static analysis, the six actuation moments (revolute joints) or forces (prismatic joints) are known, and the six values which define the external wrench $\hat{w}_{ext}$ are unknown. Six equations are obtained from (3.45) and/or (3.46) from which the six componenets of the external wrench $\hat{w}_{ext}$ can be calculated.

It is interesting to consider a manipulator with six joints for the case where gravity can be neglected. For this case

$$^{i-1}\hat{w}^i = -\hat{w}_{ext}, \quad i = 1 \ldots 6. \tag{3.51}$$

For discussion, assume all six joints are revolute joints. Equation (3.46) may be rewritten as

$$
\begin{aligned}
_{i-1}\tau_i &= {}^{i-1}S^i \cdot \left({}^{i-1}m^i\right) - {}^{i-1}S^i \cdot \left(r_i \times {}^{i-1}f^i\right) \\
&= {}^{i-1}S^i \cdot \left({}^{i-1}m^i\right) + \left(r_i \times {}^{i-1}S^i\right) \cdot {}^{i-1}f^i
\end{aligned} \tag{3.52}
$$

for $i = 1 \ldots 6$. Recognizing that $r_i \times {}^{i-1}Si = {}^{i-1}S^i_{OL}$, i.e., the moment of the line along the axis of rotation, (3.52) may be written as

$$_{i-1}\tau_i = {}^{i-1}S^i \cdot {}^{i-1}m^i + {}^{n-1}S^n_{OL} \cdot {}^{i-1}f^i, \quad i = 1 \ldots 6. \tag{3.53}$$

The external wrench will be written as $\hat{w}_{ext} = \{f_{ext}; m_{ext}\}$. Since for this case $^{i-1}\hat{w}^i = -\hat{w}_{ext}$, (3.53) may be written as

$$_{i-1}\tau_i = -\left({}^{i-1}S^i \cdot m_{ext} + {}^{n-1}S^n_{OL} \cdot f_{ext}\right), \quad i = 1 \ldots 6. \tag{3.54}$$

The six equations associated with (3.54) can be written in matrix format as

$$\tau = -J^T \begin{bmatrix} m_{ext} \\ f_{ext} \end{bmatrix}, \tag{3.55}$$

where

$$\tau = \begin{bmatrix} _0\tau_1 \\ _1\tau_2 \\ _2\tau_3 \\ _3\tau_4 \\ _4\tau_5 \\ _5\tau_6 \end{bmatrix}, \tag{3.56}$$

and J is a 6×6 matrix written as

$$J = \begin{bmatrix} {}^0S^1 & {}^1S^2 & {}^2S^3 & {}^3S^4 & {}^4S^5 & {}^5S^6 \\ {}^0S^1_{OL} & {}^1S^2_{OL} & {}^2S^3_{OL} & {}^3S^4_{OL} & {}^4S^5_{OL} & {}^5S^6_{OL} \end{bmatrix}. \tag{3.57}$$

Note that the columns of J are the Plücker coordinates of the lines associated with the six joint axes of the manipulator. For the case of no gravity, equation (3.55) can readily be used for either the forward or reverse static analysis. The matrix J is calculated in terms of the manipulator's geometry, i.e., the constant mechanism parameters and the values for the joint angles. For the forward static analysis, the vector τ is given and f_{ext} and m_{ext} are calculated. For the reverse static analysis, f_{ext} and m_{ext} are given and the vector τ is calculated.

If one of the joints of the manipulator is a prismatic joint, the appropriate column of J is replaced by the Plücker coordinates of a line at infinity, i.e., $\{0; S\}$, where S is a unit vector in the direction of the translation. The corresponding row of τ is replaced by the joint force of the prismatic joint. The representation of a translation by a line at infinity is discussed in Chapter 4.

3.8.1   Sample Problem

The constant mechanism parameters of a six axis serial manipulator comprised of seven bodies interconnected by six revolute joints are given in Table 3.1. The ground link is numbered as body 0, and the last moving link is numbered as body 6. A standard coordinate system is attached to each moving link in the manner described in Section 2.12. The fixed coordinate system is attached to body 0 according to the definition presented in Section 2.11.5 and as shown in Figure 2.24.

The value for the offset $_5S_6$ was chosen as $_5S_6 = 0.12$ m, and a line has been etched into body 6 (the last moving body) to establish the direction for the vector a_6 in body 6. Table 3.2 shows the mass of each link and the coordinates of the center of mass point as measured in the coordinate system attached to that link. The direction of gravity is parallel to the $-z$ direction and, thus, $S_{grav} = [0, 0, -1]^T$. At this instant, the joint angle parameters are measured as shown in Table 3.3. An external wrench, $\hat{w}_{ext}$, is applied to body 6, where

$$\hat{w}_{ext} = \begin{bmatrix} f_{ext} \\ m_{Oext} \end{bmatrix} = \begin{bmatrix} 8 \\ 6 \\ 7 \\ 1.66 \\ 4.24 \\ -0.172 \end{bmatrix}.$$

The units of the first three terms of $\hat{w}$, i.e., f_{ext}, are N, and the units of the last three terms of $\hat{w}$, i.e., m_{Oext}, are N–m. The objective is to determine the joint torques at the six revolute joints that will hold the manipulator in static equilibrium.

The first step is to obtain the 4×4 transformation matrices that relate coordinate system $i + 1$ to coordinate system i. The matrix ^0_1T is defined by (2.80). The matrices ^i_jT, where $j = i + 1$ and $i = 1 \ldots 5$, are defined by (2.79). From these equations, the

Table 3.1. Constant mechanism parameters

Link Length, m	Twist Angle, deg	Joint Offset, m
$a_1 = 0$	$\alpha_1 = 90$	
$a_2 = 0.23$	$\alpha_2 = 0$	$_1S_2 = 0.15$
$a_3 = 0.15$	$\alpha_3 = 270$	$_2S_3 = 0.11$
$a_4 = 0.10$	$\alpha_4 = 90$	$_3S_4 = 0.18$
$a_5 = 0.08$	$\alpha_5 = 90$	$_4S_5 = 0.08$

Table 3.2. Link mass properties

Body #	mass, kg	center of mass point, m
1	$m_1 = 9.3$	$^1\boldsymbol{P}_{m1} = \begin{bmatrix} 0.05 \\ 0 \\ 0 \end{bmatrix}$
2	$m_2 = 5.3$	$^2\boldsymbol{P}_{m2} = \begin{bmatrix} 0.2 \\ 0 \\ -0.1 \end{bmatrix}$
3	$m_3 = 3.5$	$^3\boldsymbol{P}_{m3} = \begin{bmatrix} 0.05 \\ 0.1 \\ 0 \end{bmatrix}$
4	$m_4 = 2.6$	$^4\boldsymbol{P}_{m4} = \begin{bmatrix} 0.15 \\ 0.05 \\ 0 \end{bmatrix}$
5	$m_5 = 1.8$	$^5\boldsymbol{P}_{m5} = \begin{bmatrix} 0.04 \\ 0.03 \\ 0.03 \end{bmatrix}$
6	$m_6 = 1.2$	$^6\boldsymbol{P}_{m6} = \begin{bmatrix} 0.02 \\ -0.03 \\ 0.05 \end{bmatrix}$

Table 3.3. Joint
parameter values

Joint Angle, deg
$_0\phi_1 = 225$
$_1\theta_2 = 150$
$_2\theta_3 = -60$
$_3\theta_4 = 45$
$_4\theta_5 = 60$
$_5\theta_6 = -30$

transformation matrices that relate each of the link coordinate systems with ground,
body 0, are evaluated as

$$_1^0\boldsymbol{T} = \begin{bmatrix} -0.7071 & 0.7071 & 0 & 0 \\ -0.7071 & -0.7071 & 0 & 0 \\ 0 & 0 & 1 & 0 \\ 0 & 0 & 0 & 1 \end{bmatrix} \tag{3.58}$$

$$_2^0\boldsymbol{T} = {}_1^0\boldsymbol{T}\,{}_2^1\boldsymbol{T} = \begin{bmatrix} 0.6124 & 0.3536 & -0.7071 & -0.1061 \\ 0.6124 & 0.3536 & 0.7071 & 0.1061 \\ 0.5 & -0.8660 & 0 & 0 \\ 0 & 0 & 0 & 1 \end{bmatrix} \tag{3.59}$$

$$\frac{0}{3}T = \frac{0}{2}T\,\frac{2}{3}T = \begin{bmatrix} 0 & 0.7071 & -0.7071 & -0.0430 \\ 0 & 0.7071 & 0.7071 & 0.3247 \\ 1 & 0 & 1 & 0.1150 \\ 0 & 0 & 0 & 1 \end{bmatrix} \tag{3.60}$$

$$\frac{0}{4}T = \frac{0}{3}T\,\frac{3}{4}T = \begin{bmatrix} 0.5000 & 0.5000 & 0.7071 & 0.0843 \\ -0.5000 & -0.5000 & 0.7071 & 0.4520 \\ 0.7071 & -0.7071 & 0 & 0.2650 \\ 0 & 0 & 0 & 1 \end{bmatrix} \tag{3.61}$$

$$\frac{0}{5}T = \frac{0}{4}T\,\frac{4}{5}T = \begin{bmatrix} 0.8624 & -0.0795 & -0.5000 & 0.0943 \\ 0.3624 & 0.7866 & 0.5000 & 0.4420 \\ 0.3536 & -0.6124 & 0.7071 & 0.3923 \\ 0 & 0 & 0 & 1 \end{bmatrix} \tag{3.62}$$

$$\frac{0}{6}T = \frac{0}{5}T\,\frac{5}{6}T = \begin{bmatrix} 0.9968 & -0.0018 & 0.0795 & 0.1728 \\ 0.0638 & 0.6142 & -0.7866 & 0.3766 \\ -0.0474 & 0.7892 & 0.6124 & 0.4941 \\ 0 & 0 & 0 & 1 \end{bmatrix}. \tag{3.63}$$

Note that the upper left 3×3 portion of these matrices are dimensionless, while the first three elements of the fourth column have units of meters.

The next step of the analysis is to determine the coordinates of the lines through each of the link center of mass points in the direction of gravity. These line coordinates will be written in the fixed, body 0, coordinate system. The directions of these lines are all parallel to $[0, 0, -1]^T$. To determine the moments of these lines, it is necessary to calculate the coordinates of the center of mass point of each link in terms of the fixed coordinate system. The center of mass point coordinates are determined from

$$^0 r_{iCM} = \,^0_i T \,^i r_{iCM}, \; i = 1 \dots 6. \tag{3.64}$$

These coordinates are evaluated based on the information in Table 3.2 and (3.58) through (3.63) as

$$^0 r_{1CM} = \begin{bmatrix} -0.0354 \\ -0.0354 \\ 0 \end{bmatrix}, \; ^0 r_{2CM} = \begin{bmatrix} 0.0871 \\ 0.1578 \\ 0.1000 \end{bmatrix}, \; ^0 r_{3CM} = \begin{bmatrix} 0.0277 \\ 0.3954 \\ 0.1650 \end{bmatrix},$$

$$^0 r_{4CM} = \begin{bmatrix} 0.1843 \\ 0.3520 \\ 0.3357 \end{bmatrix}, \; ^0 r_{5CM} = \begin{bmatrix} 0.1114 \\ 0.4951 \\ 0.4093 \end{bmatrix}, \; ^0 r_{6CM} = \begin{bmatrix} 0.1968 \\ 0.3201 \\ 0.5000 \end{bmatrix}, \tag{3.65}$$

where the coordinates have units of meters. The coordinates of the lines in the direction of gravity through the center of mass point are now obtained from (3.44) as

$$\$_{1Gravity} = \begin{bmatrix} 0 \\ 0 \\ -1 \\ 0.0354 \\ -0.0354 \\ 0 \end{bmatrix}, \$_{2Gravity} = \begin{bmatrix} 0 \\ 0 \\ -1 \\ -0.1578 \\ 0.0871 \\ 0 \end{bmatrix}, \$_{3Gravity} = \begin{bmatrix} 0 \\ 0 \\ -1 \\ -0.3954 \\ 0.0277 \\ 0 \end{bmatrix},$$

$$\$_{4Gravity} = \begin{bmatrix} 0 \\ 0 \\ -1 \\ -0.3520 \\ 0.1843 \\ 0 \end{bmatrix}, \$_{5Gravity} = \begin{bmatrix} 0 \\ 0 \\ -1 \\ -0.4951 \\ 0.1114 \\ 0 \end{bmatrix}, \$_{6Gravity} = \begin{bmatrix} 0 \\ 0 \\ -1 \\ -0.3201 \\ 0.1968 \\ 0 \end{bmatrix},$$

$$(3.66)$$

where the first three terms are dimensionless, and the last three have units of meters.

The wrench that body i is applying on body $i + 1$, $i = 0 \dots 5$ is determined from (3.49) as

$$^5\hat{w}^6 = \begin{bmatrix} -8 \\ -6 \\ 4.7720 \\ 2.1082 \\ -6.5563 \\ 0.1720 \end{bmatrix}, {}^4\hat{w}^5 = \begin{bmatrix} -8 \\ -6 \\ 22.4300 \\ 10.8500 \\ -8.5232 \\ 0.1720 \end{bmatrix}, {}^3\hat{w}^4 = \begin{bmatrix} -8 \\ -6 \\ 47.9360 \\ 19.8274 \\ -13.2234 \\ 0.1720 \end{bmatrix},$$

$$^2\hat{w}^3 = \begin{bmatrix} -8 \\ -6 \\ 82.2710 \\ 33.4036 \\ -14.1748 \\ 0.1720 \end{bmatrix}, {}^1\hat{w}^2 = \begin{bmatrix} -8 \\ -6 \\ 134.2640 \\ 41.6097 \\ -18.7044 \\ 0.1720 \end{bmatrix}, {}^0\hat{w}^1 = \begin{bmatrix} -8 \\ -6 \\ 225.4970 \\ 38.3841 \\ -15.4788 \\ 0.1720 \end{bmatrix}, (3.67)$$

where the first three components have units of N, and the last three components have units of N–m.

Equation (3.46) is used to determine the individual joint torques. In this equation, the term r_n represents any point on the line of action of the revolute joint being considered. The origin point of that link's standard coordinate system is on the line of action of the joint, and the coordinates of this point can be obtained from the first three elements of the fourth column of the matrices (3.58)–(3.63). For this numerical case, the joint torques are calculated as

$$_0\tau_1 = 0.1720, \ _1\tau_2 = -42.6485, \ _2\tau_3 = -16.1174,$$
$$_3\tau_4 = -7.4188, \ _4\tau_5 = -2.9052, \ _5\tau_6 = 0.0842 \text{ N–m.}$$

3.8.2   Sample Problem

The previous problem will be repeated, ignoring the mass of the links. Equation (3.55) can be used to solve this problem. The matrix J is defined in (3.57), and its columns are the coordinates of the lines along the six revolute axes of the manipulator, all written in the body 0 coordinate system. The first three terms in the third column of (3.58)–(3.63) give the direction of the line along each revolute joint. The first three terms in the fourth column of these matrices provide a point on the joint axis. The cross product of the coordinates of this point with the direction of the line gives the moment of the line with respect to the origin of the fixed coordinate system. The matrix J is calculated as

$$
J = \begin{bmatrix}
0 & -0.7071 & -0.7071 & 0.7071 & -0.5 & 0.0795 \\
0 & 0.7071 & 0.7071 & 0.7071 & 0.5 & -0.7866 \\
1 & 0 & 0 & 0 & 0.7071 & 0.6124 \\
0 & 0 & -0.0813 & -0.1874 & 0.1164 & 0.6192 \\
0 & 0 & -0.8132 & 0.1874 & -0.2628 & -0.0666 \\
0 & 0 & 0.1992 & -0.2600 & 0.2681 & -0.1658
\end{bmatrix}.
\tag{3.68}
$$

The individual joint torques are calculated from (3.55) as

$$
{}_0\tau_1 = 0.1720, \quad {}_1\tau_2 = -1.8243, \quad {}_2\tau_3 = -2.0802,
$$
$$
{}_3\tau_4 = -1.9772, \quad {}_4\tau_5 = -2.3995, \quad {}_5\tau_6 = -0.0849 \text{ N–m.}
$$

3.9   The Resultant of a Pair of Wrenches Acting Upon a Rigid Body

Consider a given pair of wrenches that are acting upon a rigid body (see Figure 3.13). The perpendicular distance between the lines of action of these wrenches is a_{12}, and a_{12} is a unit vector along the line that is mutually perpendicular to the lines of action of the two wrenches. The angle between f_1 and f_2 is labeled as α_{12} and is measured in a right-hand sense about a_{12}. For convenience, and without loss of generality, the reference point O is chosen at the intersection of the line of action of the first wrench and the mutually perpendicular line between the two wrenches. The coordinates of the two given wrenches may, thus, be written as $\{f_1; h_1 f_1\}$ and $\{f_2; a_{12} a_{12} \times f_2 + h_2 f_2\}$. The vector a_{12} may be determined from

$$
a_{12} = \frac{f_1 \times f_2}{|f_1 \times f_2|},
\tag{3.69}
$$

and the angle α_{12} between the two wrenches may be uniquely determined from the two equations

$$
\cos \alpha_{12} = \frac{f_1 \cdot f_2}{|f_1||f_2|},
\tag{3.70}
$$

$$
\sin \alpha_{12} = \frac{f_1 \times f_2 \cdot a_{12}}{|f_1||f_2|}.
\tag{3.71}
$$

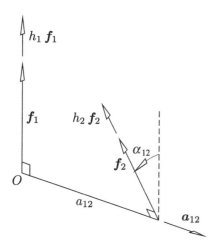

Figure 3.13 A pair of wrenches acting on a rigid body

The coordinates of the resultant wrench $\{f; m_O\}$ can be expressed as

$$\begin{bmatrix} f \\ m_O \end{bmatrix} = \begin{bmatrix} f_1 \\ h_1 f_1 \end{bmatrix} + \begin{bmatrix} f_2 \\ a_{12} a_{12} \times f_2 + h_2 f_2 \end{bmatrix} \qquad (3.72)$$

and, therefore,

$$f = f_1 + f_2 \qquad (3.73)$$

and

$$m_O = h_1 f_1 + a_{12} a_{12} \times f_2 + h_2 f_2. \qquad (3.74)$$

The pitch of the resultant may be written as

$$h = \frac{f \cdot m_O}{f \cdot f}, \qquad (3.75)$$

and the Plücker coordinates of the line along the resultant are $\frac{1}{f}\{f; m_O - hf\}$, where $f = |f|$.

The Plücker coordinates of the line that is perpendicular to the line of action of the original two wrenches may be written as $\{a_{12}; 0\}$. Evaluating the mutual moment of this line with the line along the resultant gives

$$a_{12} \cdot \frac{1}{f}(m_O - hf) = a_{12} \cdot \frac{1}{f} \left[(h_1 f_1 + a_{12} a_{12} \times f_2 + h_2 f_2) - a_{12} \cdot hf \right]. \qquad (3.76)$$

Upon multiplying (3.76) by f and recognizing that a_{12} is perpendicular to f_1 and f_2, equation (3.76) reduces to

$$a_{12} \cdot (m_O - hf) = -a_{12} \cdot hf. \qquad (3.77)$$

Upon substituting (3.73) into (3.77), it is apparent that the right side of (3.77) vanishes. Since the mutual moment of the line of action along the resultant with the

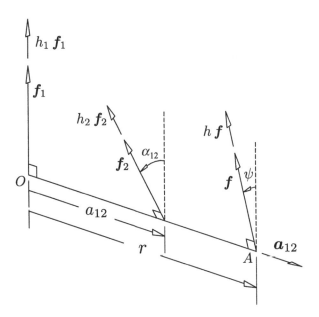

Figure 3.14 Resultant of a pair of wrenches

line that is mutually perpendicular to the two given wrenches is zero, then these two lines must either be parallel or they must intersect. It is apparent from (3.69) and (3.73) that the directions of the lines are not parallel and, thus, the line of action along the resultant intersects and is perpendicular to the line $\{a_{12}; 0\}$, as shown in Figure 3.14. In this figure, the line of action of the resultant passes through a point A. The distance between points O and A is labeled r, and the angle between the direction of the resultant wrench and the first wrench is labeled as ψ.

The resultant of the two original wrenches may be determined graphically. Firstly the force triangle is drawn in Figure 3.15(a), which yields f and ψ. Secondly, a moment diagram is drawn in Figure 3.15(b) by drawing the vector $m_{O1} = h_1 f_1$ parallel to f_1. Then, the vectors $a_{12}a_{12} \times f_2$ and $h_2 f_2$ are respectively drawn perpendicular to and parallel to f_2, which yields the resultant m_{O2}. The resultant m_O is the vector addition of m_{O1} and m_{O2}. Finally, m_O can be decomposed into a vector $h f$ parallel to f and a vector $r\,a_{12} \times f$ perpendicular to $h f$. From the two polygons, $h = \frac{|h f|}{|f|}$ and $r = \frac{|r\,a_{12} \times f|}{|f|}$. Hence, f, h, ψ, and r are determined, which completely specify the resultant wrench.

Alternatively, applying the cosine and sine laws to the force triangle of Figure 3.15(a) yields

$$f^2 = f_1^2 + f_2^2 + 2 f_1 f_2 \cos \alpha_{12} \tag{3.78}$$

and

$$\frac{f}{\sin \alpha_{12}} = \frac{f_2}{\sin \psi}, \tag{3.79}$$

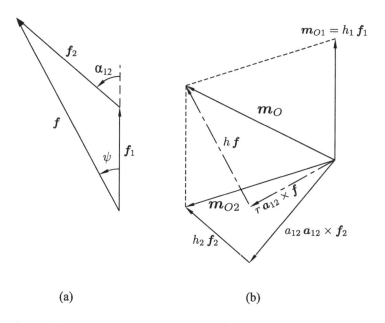

Figure 3.15 Force and moment polygons: (a) force polygon and (b) moment polygon

$$\frac{f}{\sin \alpha_{12}} = \frac{f_1}{\sin(\alpha_{12} - \psi)}. \tag{3.80}$$

Equation (3.80) may be expanded as

$$f_1 \sin \alpha_{12} = f(\sin \alpha_{12} \cos \psi - \cos \alpha_{12} \sin \psi). \tag{3.81}$$

The value for the magnitude for the resultant, i.e., f, may be computed from (3.78). The sine of ψ may then be determined from (3.79) and the cosine of ψ from (3.81), which yields the unique value for ψ. Further, the pitch h is given by

$$h = \frac{\boldsymbol{f} \cdot \boldsymbol{m}_O}{\boldsymbol{f} \cdot \boldsymbol{f}} = \frac{\boldsymbol{f} \cdot (h_1 \boldsymbol{f}_1 + a_{12} \boldsymbol{a}_{12} \times \boldsymbol{f}_2 + h_2 \boldsymbol{f}_2)}{f^2}. \tag{3.82}$$

Now from the force triangle (Figure 3.15(a))

$$\boldsymbol{f} \cdot \boldsymbol{f}_1 = f \, f_1 \cos \psi, \tag{3.83}$$

$$\boldsymbol{f} \cdot \boldsymbol{f}_2 = f \, f_2 \cos(\alpha_{12} - \psi). \tag{3.84}$$

Also,

$$\boldsymbol{f} \cdot (a_{12} \boldsymbol{a}_{12} \times \boldsymbol{f}_2) = (\boldsymbol{f}_1 + \boldsymbol{f}_2) \cdot (a_{12} \boldsymbol{a}_{12} \times \boldsymbol{f}_2), \tag{3.85}$$

which simplifies to

$$\boldsymbol{f} \cdot (a_{12} \boldsymbol{a}_{12} \times \boldsymbol{f}_2) = -a_{12} \boldsymbol{a}_{12} \cdot \boldsymbol{f}_1 \times \boldsymbol{f}_2. \tag{3.86}$$

Substituting $\boldsymbol{f}_1 \times \boldsymbol{f}_2 = f_1 f_2 \sin \alpha_{12} \boldsymbol{a}_{12}$ into (3.86) yields

$$\boldsymbol{f} \cdot (a_{12} \boldsymbol{a}_{12} \times \boldsymbol{f}_2) = -f_1 f_2 a_{12} \sin \alpha_{12}. \tag{3.87}$$

Substituting (3.83), (3.84), and (3.87) into (3.82) gives

$$h = \frac{f_1}{f} h_1 \cos \psi + \frac{f_2}{f} h_2 \cos(\alpha_{12} - \psi) - \frac{f_1 f_2}{f^2} a_{12} \sin \alpha_{12}. \tag{3.88}$$

An expression for the distance r is obtained from the mutual moment of the lines along the axes of the first given wrench and the resultant. The Plücker coordinates of these lines are $\frac{1}{f_1}\{f_1; 0\}$ and $\frac{1}{f}\{f; m_O - hf\}$, and from (1.140) the mutual moment may be written as

$$\frac{1}{f_1 f} f_1 \cdot (m_O - h f) = -r \sin \psi. \tag{3.89}$$

Multiplying (3.89) by $f_1 f$ and utilizing (3.74) gives

$$f_1 \cdot m_O = f_1 \cdot (h_1 f_1 + a_{12} a_{12} \times f_2 + h_2 f_2), \tag{3.90}$$

which can be written as

$$f_1 \cdot m_O = f_1^2 h_1 - f_1 f_2 a_{12} \sin \alpha_{12} + f_1 f_2 h_2 \cos \alpha_{12}. \tag{3.91}$$

Substituting (3.91) together with $f_1 \cdot f = f_1 \cdot (f_1 + f_2) = f_1^2 + f_1 f_2 \cos \alpha_{12}$ into the left side of (3.89) and then rearranging yields

$$r \sin \psi = \frac{f_1}{f}(h - h_1) + \frac{f_2}{f}\{(h - h_2) \cos \alpha_{12} + a_{12} \sin \alpha_{12}\}. \tag{3.92}$$

Thus, the values for f, ψ, h, and r for the resultant of the two original wrenches may be determined from equations (3.78), (3.79) and (3.81), (3.88), and (3.92), respectively.

3.10   The Cylindroid

Consider again two wrenches acting on a rigid body, as shown in Figure 3.13, where for convenience the reference point O is chosen on the axis of the first wrench at the intersection with the line mutually perpendicular to the lines of action of the two wrenches. As the ratio of the magnitudes of the wrenches is varied, i.e., the ratio $\frac{f_1}{f_2}$, the axis of the resultant wrench will move on a ruled surface, the generators of which intersect the mutually perpendicular line of the axes of the original wrenches. Ball, at the suggestion of Cayley, called this surface the cylindroid. It also goes by Plücker's conoid.

To draw the cylindroid, it is necessary to determine values for ψ, f, r, and h of the resultant as the ratio $\frac{f_1}{f_2}$ varies from $-\infty$ to $+\infty$. From the force polygon shown in Figure 3.15(a), it is apparent that when f_2 is much smaller than f_1, and the ratio $\frac{f_1}{f_2}$ approaches $+\infty$, ψ approaches 0. If the ratio $\frac{f_1}{f_2}$ approaches $-\infty$ by f_1 being a negative value and f_2 approaching 0, then ψ approaches π. As the ratio $\frac{f_1}{f_2}$ varies between $-\infty$ and $+\infty$, the angle ψ will vary between 0 and 2π, with one unique value of ψ corresponding to each value of $\frac{f_1}{f_2}$. It is always possible to determine the value of the ratio $\frac{f_1}{f_2}$ for a given value of ψ. Equating the right sides of (3.79) and (3.80) gives

$$\frac{f_2}{\sin \psi} = \frac{f_1}{\sin(\alpha_{12} - \psi)}. \tag{3.93}$$

This can be rearranged as

$$\frac{f_1}{f_2} = \frac{\sin(\alpha_{12} - \psi)}{\sin \psi}. \tag{3.94}$$

Since the angle of the resultant, ψ, varies in the range of 0 to 2π, the cylindroid can be generated by determining values for r and h as functions of ψ. The following pair of equations for the location and pitch of the screws on the cylindroid are derived in Appendix A:

$$r = R(\sin \sigma + \sin(2\psi - \sigma)), \tag{3.95}$$

$$h = h_1 + R(\cos \sigma - \cos(2\psi - \sigma)), \tag{3.96}$$

where

$$R = \sqrt{\frac{a_{12}^2 + (h_2 - h_1)^2}{4 \sin^2 \alpha_{12}}}, \tag{3.97}$$

$$\sin \sigma = \frac{a_{12} - (h_2 - h_1) \cot \alpha_{12}}{2R}, \tag{3.98}$$

$$\cos \sigma = \frac{(h_2 - h_1) + a_{12} \cot \alpha_{12}}{2R}. \tag{3.99}$$

Figure 3.16 shows plots of r and h vs ψ for the case where the pitches of the two original screws are given as $h_1 = 4$ m and $h_2 = -3$ m, and the angle and distance between the screws are given as $\alpha_{12} = 40°$ and $a_{12} = 2$ m.

A cylindroid is illustrated in Figures 3.17 and 3.18 for the case where $\hat{w}_1 = f_1\{1, 0, 0; 10, 0, 0\}$ and $\hat{w}_2 = f_2\{0.7660, 0.6428, 0; 2.2716, 9.7386, 0\}$, where the first three components of each screw are dimensionless, and the last three have units of meters. For this example problem, the line of action of the first screw is along the X axis, and the unique perpendicular line between the lines along the two given screws is along the Z axis. The pitches of the two given wrenches were determined to be $h_1 = 10$ m and $h_2 = 8$ m. The perpendicular distance between the lines of action of the wrenches was determined to be $a_{12} = 6$ m, and the angle between the directions of the lines of action was evaluated as $\alpha_{12} = 40°$ measured with respect to the vector a_{12}, whose direction was chosen as parallel to $S_1 \times S_2$. The value of R was determined from (3.97) as $R = 4.9196$ m, and the sine and cosine of σ were determined from (3.98) and (3.99) as $\sin \sigma = 0.8520$ and $\cos \sigma = 0.5235$. The position r, and pitch, h, are then determined from (3.95) and (3.96) for each value of ψ. The resultant wrenches are drawn as ψ varies from 0 to 2π. The pitch of each resultant is indicated by the length of the cylinders that are drawn to scale in units of meters.

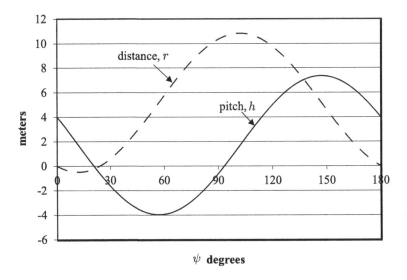

Figure 3.16 Resultant pitch and distance from first screw for case where $h_1 = 4$ m, $h_2 = -3$ m, $\alpha_{12} = 40°$, and $a_{12} = 2$ m

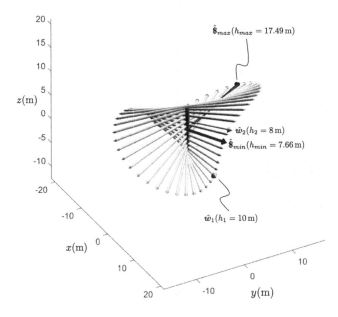

Figure 3.17 Cylindroid

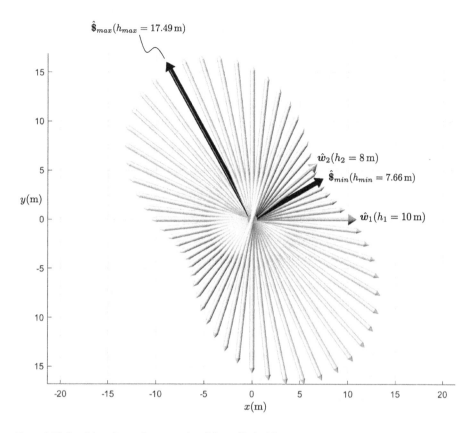

Figure 3.18 Looking down the a_{12} axis of the cylindroid

The length of the cylindroid (see Figure 3.19) will now be determined from (3.95). The maximum value of r, i.e., $r_{\max}$, will occur when $\sin(2\psi - \sigma) = 1$ or $2\psi - \sigma = \pi/2$. The value of ψ corresponding to $r_{\max}$ can be determined as

$$\psi = \frac{\frac{\pi}{2} + \sigma}{2} \qquad (3.100)$$

and

$$r_{\max} = R(\sin\sigma + 1). \qquad (3.101)$$

The minimum value of r, i.e., $r_{\min}$, will occur when $\sin(2\psi - \sigma) = -1$ or $2\psi - \sigma = 3\pi/2$. The value of ψ corresponding to $r_{\min}$ can be determined as

$$\psi = \frac{\frac{3\pi}{2} + \sigma}{2} \qquad (3.102)$$

and

$$r_{\min} = R(\sin\sigma - 1). \qquad (3.103)$$

From (3.101) and (3.103), the length of the cylindroid is given by

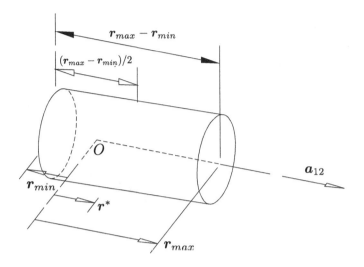

Figure 3.19 Length of cylindroid (note $r_{min} < 0$)

$$r_{max} - r_{min} = 2R. \tag{3.104}$$

From (3.100) and (3.102), the angle between the screws at the extreme ends of the cylindroid is $\pi/2$.

From (3.96), the maximum pitch, h_{max} will occur when $\cos(2\psi - \sigma) = -1$ or $2\psi - \sigma = \pi$. The value of ψ corresponding to h_{max} can be determined as

$$\psi = \frac{\pi + \sigma}{2} \tag{3.105}$$

and

$$h_{max} = h_1 + R(\cos \sigma + 1). \tag{3.106}$$

From (3.96), the minimum pitch, h_{min}, will occur when $\cos(2\psi - \sigma) = 1$ or $2\psi - \sigma = 0$. The value of ψ corresponding to h_{min} can be determined as

$$\psi = \frac{\sigma}{2} \tag{3.107}$$

and

$$h_{min} = h_1 + R(\cos \sigma - 1). \tag{3.108}$$

From (3.105) and (3.107), the angle between the screws of maximum and minimum pitch is, therefore, $\pi/2$. The locations of the screws of maximum and minimum pitch can be determined by substituting (3.105) and (3.107) into (3.95). The same value of r is determined for both of these screws (named r^*) and is evaluated as

$$r^* = R \sin \sigma. \tag{3.109}$$

The screws with maximum and minimum pitch will be referred to as the principal screws of the cylindroid, and the term r^* represents the distance of the principal screws from the first given screw along the mutual perpendicular a_{12}. Adding (3.101) and (3.103) and dividing by two gives

$$\frac{r_{max} + r_{min}}{2} = R \sin \sigma = r^*, \tag{3.110}$$

and r^* is, thus, the distance to the midpoint of the cylindroid. Clearly, the principal screws of the cylindroid intersect at an angle of $\pi/2$ at the midpoint of the cylindroid.

3.11 Circular Representation of the Cylindroid

The property that the axes of the principal screws intersect at right angles will now be used in a relatively simple derivation for the pitches and location of the screws on the cylindroid. It is assumed that the screws of maximum and minimum pitch, $\$_1$ and $\$_2$, lie on the x and y axes and have pitches equal to h_1 and h_2, respectively ($h_1 > h_2$), as shown in Figure 3.20. The coordinates for the screws are $\hat{s}_1 = \{1, 0, 0; h_1, 0, 0\}$ and $\hat{s}_2 = \{0, 1, 0; 0, h_2, 0\}$, respectively. Any screw $\$$ on the cylindroid is a linear combination of $\$_1$ and $\$_2$ and, hence,

$$\lambda \hat{s} = \lambda_1 \hat{s}_1 + \hat{s}_2, \tag{3.111}$$

which may be written as

$$\lambda \hat{s} = \{\lambda_1, 1, 0; \lambda_1 h_1, h_2, 0\}. \tag{3.112}$$

The coordinates of the resultant may be written as $\{S; S_O\}$, where $S = \lambda_1 i + j$ and $S_O = \lambda_1 h_1 i + h_2 j$. The magnitude of $\$$ is given by

$$\lambda = \sqrt{1 + \lambda_1^2}, \tag{3.113}$$

and the orientation of the resultant, θ, (see Figure 3.20) is given by

$$\cos \theta = \frac{\lambda_1}{\sqrt{1 + \lambda_1^2}}, \quad \sin \theta = \frac{1}{\sqrt{1 + \lambda_1^2}}. \tag{3.114}$$

From (3.112), the pitch of the resultant is given by

$$h = \frac{S \cdot S_O}{S \cdot S} = \frac{\lambda_1^2 h_1 + h_2}{1 + \lambda_1^2} = h_1 \cos^2 \theta + h_2 \sin^2 \theta. \tag{3.115}$$

Substituting $\cos 2\theta = 2 \cos^2 \theta - 1 = 1 - 2 \sin^2 \theta$ yields the equation for the pitch cone

$$h = \frac{1}{2}(h_1 + h_2) + \frac{1}{2}(h_1 - h_2) \cos 2\theta. \tag{3.116}$$

Now the equation for the line of action of the resultant $\$$ is given by

$$r \times S = S_O - hS \tag{3.117}$$

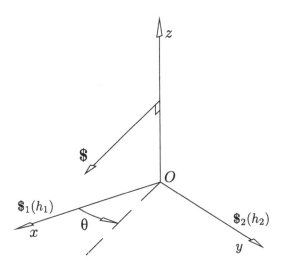

Figure 3.20 Principal screws $\$_1$ and $\$_2$ and resultant $\$$

or

$$zk \times S = S_O - hS. \tag{3.118}$$

Forming the cross product of (3.118) with S yields

$$S \times (zk \times S) = S \times (S_O - hS). \tag{3.119}$$

Expanding the left and right sides of (3.119) gives

$$zk (S \cdot S) - S (S \cdot zk) = S \times S_O - S \times hS. \tag{3.120}$$

Substituting $S \cdot k = 0$ and $S \times S = \mathbf{0}$ into (3.120) gives

$$zk (S \cdot S) = S \times S_O, \tag{3.121}$$

and substituting for S and S_O and rearranging gives

$$zk = \frac{1}{1 + \lambda_1^2} \begin{vmatrix} i & j & k \\ \lambda_1 & 1 & 0 \\ \lambda_1 h_1 & h_2 & 0 \end{vmatrix}. \tag{3.122}$$

Expanding the right side of (3.122) yields

$$zk = \frac{(h_2 - h_1)\lambda_1}{1 + \lambda_1^2} k. \tag{3.123}$$

From (3.123),

$$z = (h_2 - h_1)\frac{\lambda_1}{1 + \lambda_1^2}. \tag{3.124}$$

From (3.114), $\frac{\lambda_1}{1+\lambda_1^2} = \sin \theta \cos \theta$, and substituting this into (3.124) gives

$$z = (h_2 - h_1) \sin \theta \cos \theta \tag{3.125}$$

and upon substituting $\sin 2\theta = 2 \sin \theta \cos \theta$,

$$z = \frac{1}{2}(h_2 - h_1) \sin 2\theta. \tag{3.126}$$

Equation (3.116) can be rearranged into the form

$$h - \frac{1}{2}(h_1 + h_2) = \frac{1}{2}(h_1 - h_2) \cos 2\theta. \tag{3.127}$$

Squaring both sides of (3.126) and (3.127), adding the results, and recognizing that $(\sin 2\theta)^2 + (\cos 2\theta)^2 = 1$, yields

$$z^2 + \left(h - \frac{1}{2}(h_1 + h_2)\right)^2 = \left(\frac{1}{2}(h_1 - h_2)\right)^2. \tag{3.128}$$

Equation (3.128) is clearly the equation of a circle in the zh plane, with it's center located at the point $(0, \frac{1}{2}(h_1 + h_2))$ and a radius equal to $\frac{1}{2}(h_1 - h_2)$. This circular representation of the cylindroid is illustrated in Figure 3.21. The circle intersects the h axis at the points $(0, h_2)$ and $(0, h_1)$ since the h coordinate of the center of the circle is $\frac{1}{2}(h_1 + h_2)$ and the radius of the circle is $\frac{1}{2}(h_1 - h_2)$. The center of the circle plus the radius along the h axis gives the point $(0, h_1)$, and the center of the circle minus the radius along the h axis gives the point $(0, h_2)$.

It is apparent from (3.116) that any coordinate h is equal to the h coordinate of the center of the circle plus the radius of the circle multiplied by $\cos 2\theta$. Hence, the angle subtended at the center of the circle by the chord joining the points (z, h) and $(0, h_1)$ is 2θ. Hence, the same chord subtends an angle θ at any point on the circle.

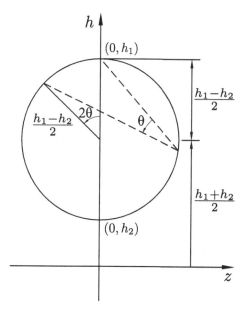

Figure 3.21   Circle representation of the cylindroid

It should be clear from Figure 3.21 that in general there are two screws of equal pitch h on the cylindroid except for the screws of minimum and maximum pitch, h_2 and h_1, which are unique for which $2\theta = \pi$ and 2π, respectively, or $\theta = \pi/2$ and π. They therefore intersect at right angles through the point $z = 0$. The cylindroid is bounded by the pair of screws with equal pitch $h = \frac{1}{2}(h_1 + h_2)$, for which $\cos 2\theta = 0$ and, thus, $\theta = \pi/4$ and $3\pi/4$. From (3.128), $z = -\frac{1}{2}(h_1 - h_2)$ and $z = \frac{1}{2}(h_1 - h_2)$.

These results support those presented in Section 3.10. It has been shown in both sections that the principal screws intersect and their directions are perpendicular to one another. Also, the screws at each end of the cylindroid have the same pitch and their directions are also perpendicular to one another.

3.12   Motor Product

Figure 3.22 shows two screws that are expressed in terms of a coordinate system with origin at point O. The directions of the lines along the screws are given by the dimensionless <u>unit vectors</u> S_1 and S_2. The vectors r_1 and r_2 are the position vectors of points on the line of action of each screw. The pitches of the screws are designated as h_1 and h_2. The position vectors and the pitches have units of length. The coordinates of the screws can thus be written as

$$\$_1 = \begin{bmatrix} S_1 \\ S_{O1} \end{bmatrix} = \begin{bmatrix} S_1 \\ r_1 \times S_1 + h_1 S_1 \end{bmatrix} \tag{3.129}$$

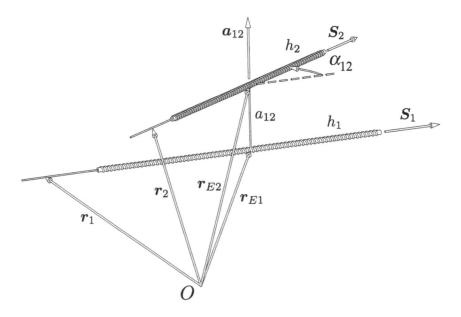

Figure 3.22 Two arbitrary screws

$$\$_2 = \begin{bmatrix} S_2 \\ S_{O2} \end{bmatrix} = \begin{bmatrix} S_2 \\ r_2 \times S_2 + h_2 S_2 \end{bmatrix}. \tag{3.130}$$

The motor product (also known as the "Lie product," "dual vector product," or "screw cross product")[4] is defined as

$$\$_1 \otimes \$_2 = \begin{bmatrix} S_1 \times S_2 \\ S_1 \times S_{O2} - S_2 \times S_{O1} \end{bmatrix}, \tag{3.131}$$

$$\$_1 \otimes \$_2 = \begin{bmatrix} S_1 \times S_2 \\ S_1 \times (r_2 \times S_2 + h_2 S_2) - S_2 \times (r_1 \times S_1 + h_1 S_1) \end{bmatrix} = \begin{bmatrix} S \\ S_O \end{bmatrix}. \tag{3.132}$$

It is apparent that the result of the motor product is another screw. It will be shown that this result represents a screw of magnitude $\sin \alpha_{12}$ with a pitch equal to $a_{12} \cot \alpha_{12} + (h_1 + h_2)$, whose line of action is along the common perpendicular to the lines of action of the two screws. The direction along the common perpendicular is defined as $a_{12} = (S_1 \times S_2) / |S_1 \times S_2|$, the distance a_{12} is defined as the perpendicular distance from the line along screw 1 to the line along screw 2 in the direction of a_{12}, and α_{12} is defined as the angle between S_1 and S_2 defined in a right-hand sense about a_{12}.

The cross product and dot product of the directions of the two screws can be written as

$$S_1 \times S_2 = \sin \alpha_{12} \, a_{12}, \tag{3.133}$$

$$S_1 \cdot S_2 = \cos \alpha_{12}. \tag{3.134}$$

Substituting (3.133) into (3.132) gives

$$S = \sin \alpha_{12} \, a_{12}. \tag{3.135}$$

The points of intersection of the mutual perpendicular line with the lines of action of the first and second screws are labeled r_{E1} and r_{E2}, as shown in Figure 3.22. These vectors may be expressed as

$$r_{E1} = r_1 + \ell_1 S_1, \tag{3.136}$$

$$r_{E2} = r_2 + \ell_2 S_2. \tag{3.137}$$

It is apparent that $r_1 \times S_1 = r_{E1} \times S_1$ and $r_2 \times S_2 = r_{E2} \times S_2$, since $\ell_i S_i \times S_i = 0$, and (3.132) can be expressed as

$$\begin{aligned} S_o &= S_1 \times (r_{E2} \times S_2 + h_2 S_2) - S_2 \times (r_{E1} \times S_1 + h_1 S_1) \\ &= S_1 \times (r_{E2} \times S_2) - S_2 \times (r_{E1} \times S_1) \\ &\quad + (h_1 + h_2)(S_1 \times S_2). \end{aligned} \tag{3.138}$$

The vector triple products $S_1 \times (r_{E2} \times S_2)$ and $S_2 \times (r_{E1} \times S_1)$ may be expressed as

$$S_1 \times (r_{E2} \times S_2) = (S_1 \cdot S_2) r_{E2} - (S_1 \cdot r_{E2}) S_2, \tag{3.139}$$

$$S_2 \times (r_{E1} \times S_1) = (S_2 \cdot S_1) r_{E1} - (S_2 \cdot r_{E1}) S_1. \tag{3.140}$$

---

[4] See also Brand (1947), and Dimentberg (1968).

Substituting (3.139) and (3.140) into (3.138) gives

$$S_o = (S_1 \cdot S_2)r_{E2} - (S_1 \cdot r_{E2})S_2 \tag{3.141}$$
$$- (S_2 \cdot S_1)r_{E1} + (S_2 \cdot r_{E1})S_1 + (h_1 + h_2)(S_1 \times S_2).$$

Substituting (3.133) and (3.134) into (3.141) yields

$$S_o = \cos\alpha_{12}(r_{E2} - r_{E1}) + (S_2 \cdot r_{E1})S_1 - (S_1 \cdot r_{E2})S_2 + \sin\alpha_{12}(h_1 + h_2)a_{12}. \tag{3.142}$$

From Figure (3.22), it is apparent that

$$r_{E2} = r_{E1} + a_{12}a_{12} \tag{3.143}$$

and, thus, (3.142) can be written as

$$S_o = (S_2 \cdot r_{E1})S_1 - (S_1 \cdot r_{E2})S_2 + [a_{12}\cos\alpha_{12} + \sin\alpha_{12}(h_1 + h_2)]a_{12}. \tag{3.144}$$

Since $S_2 \cdot a_{12} = 0$, the term $S_2 \cdot r_{E1}$ in (3.144) can be replaced by $S_2 \cdot (r_{E1} + a_{12}a_{12})$ to give

$$S_o = (S_2 \cdot (r_{E1} + a_{12}a_{12}))S_1 - (S_1 \cdot r_{E2})S_2 + [a_{12}\cos\alpha_{12} + \sin\alpha_{12}(h_1 + h_2)]a_{12}. \tag{3.145}$$

Substituting (3.143) into (3.145) yields

$$S_o = (S_2 \cdot r_{E2})S_1 - (S_1 \cdot r_{E2})S_2 + [a_{12}\cos\alpha_{12} + \sin\alpha_{12}(h_1 + h_2)]a_{12}. \tag{3.146}$$

The first two terms in (3.146) can be replaced by the vector triple product $r_{E2} \times (S_1 \times S_2)$ to give

$$S_o = r_{E2} \times (S_1 \times S_2) + [a_{12}\cos\alpha_{12} + \sin\alpha_{12}(h_1 + h_2)]a_{12}. \tag{3.147}$$

Lastly, substituting (3.133) into (3.147) gives

$$S_o = \sin\alpha_{12} r_{E2} \times a_{12} + [a_{12}\cos\alpha_{12} + \sin\alpha_{12}(h_1 + h_2)]a_{12}. \tag{3.148}$$

The results of (3.135) and (3.148) are substituted into (3.132) to yield

$$\$_1 \otimes \$_2 = \begin{bmatrix} \sin\alpha_{12}\, a_{12} \\ \sin\alpha_{12}\, r_{E2} \times a_{12} + [a_{12}\cos\alpha_{12} + \sin\alpha_{12}(h_1 + h_2)]a_{12} \end{bmatrix}, \tag{3.149}$$

which may be written as

$$\$_1 \otimes \$_2 = \sin\alpha_{12} \begin{bmatrix} a_{12} \\ r_{E2} \times a_{12} + [a_{12}\cot\alpha_{12} + (h_1 + h_2)]a_{12} \end{bmatrix}. \tag{3.150}$$

It is apparent from inspecting equation (3.150) that the motor product $\$_1 \otimes \$_2$ results in a screw of magnitude $\sin\alpha_{12}$, whose line of action is in the direction of a_{12} along a line that passes through the point r_{E2}, which means that the line of action of the resulting screw is the line that is mutually perpendicular to screws $\$_1$ and $\$_2$. Further, the pitch of the resulting screw is equal to $a_{12}\cot\alpha_{12} + (h_1 + h_2)$.

In Section 1.11 it was shown how to obtain the Plücker coordinates of the line that is mutually perpendicular to two given lines. A more straightforward approach is to use (3.131) to calculate the motor product of the two lines. The unique perpendicular line will be the line of action of the resulting screw. Note, however, that a special case occurs if the lines of actions of the two screws are parallel and, thus, the line perpendicular to the two lines of action is not unique. In this case, the motor product results in a line at infinity whose moment term is parallel to the direction perpendicular to the lines of action of the two given screws.

3.13   Problems

1. Three forces that are acting on a rigid body are given by

$$f_1 = \begin{bmatrix} 8 \\ 3 \\ -2 \end{bmatrix} \text{ N}, \quad f_2 = \begin{bmatrix} 7 \\ 0 \\ 0 \end{bmatrix} \text{ N}, \quad f_3 = \begin{bmatrix} 0 \\ 5 \\ 5 \end{bmatrix} \text{ N}.$$

These forces are acting respectively through the points

$$p_1 = \begin{bmatrix} 0 \\ 2 \\ -2 \end{bmatrix} \text{ m}, \quad p_2 = \begin{bmatrix} 4 \\ 0 \\ 8 \end{bmatrix} \text{ m}, \quad p_3 = \begin{bmatrix} 6 \\ 1 \\ -4 \end{bmatrix} \text{ m}.$$

   (a) Determine the equivalent wrench that is acting on the rigid body.
   (b) Determine the magnitude of the force along the wrench, the line of action of the wrench, and the pitch of the wrench.

2. A dyname acting on a rigid body consists of a force $[3, 2, 4]^T$ N acting through the point $[0, -2, 6]^T$ m and a couple $[5, 7, -4]^T$ Nm. Determine the equivalent wrench that is acting on the rigid body. Find the magnitude of the force along the wrench, the line of action along the wrench, and the pitch of the wrench.

3. The line of action of a wrench is given by $\hat{s} = [1, 2, -4; 0, 4, 2]$, where the first three components are dimensionless, and the last three have units of m. Determine the coordinates of the wrench if its magnitude is 20 N and its pitch is 1.5 m.

4. Figure 3.23 shows a $3 - 3$ in-parallel platform. The coordinates of points 1, 2, and 3, which are located in the base platform, are measured in the XYZ coordinate system as

$$^{XYZ}p_1 = \begin{bmatrix} 0 \\ 0 \\ 0 \end{bmatrix} \text{ m}, \quad ^{XYZ}p_2 = \begin{bmatrix} 4 \\ 0 \\ 0 \end{bmatrix} \text{ m}, \quad ^{XYZ}p_3 = \begin{bmatrix} 2 \\ 2 \\ 0 \end{bmatrix} \text{ m}.$$

The coordinates of points 4, 5, and 6, which are located in the top platform, are measured in the xyz coordinate system as

$$^{xyz}p_4 = \begin{bmatrix} 0 \\ 0 \\ 0 \end{bmatrix} \text{ m}, \quad ^{xyz}p_5 = \begin{bmatrix} 2 \\ 0 \\ 0 \end{bmatrix} \text{ m}, \quad ^{xyz}p_6 = \begin{bmatrix} 1 \\ 2 \\ 0 \end{bmatrix} \text{ m}.$$

The current position of the top platform relative to the base can be described by initially aligning the two coordinate systems and then translating the origin of the

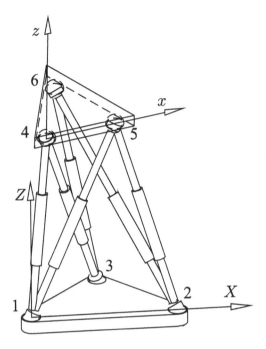

Figure 3.23 $3 - 3$ platform

xyz coordinate system to $[0.5, 0.5, 5]^T$ m. The xyz coordinate system is then rotated $10°$ about its x axis followed by $15°$ about its new z axis.

(a) Determine the coordinates of points 4, 5, and 6 in terms of the XYZ coordinate system.

(b) Determine the Plücker coordinates of the lines along the six legs of the platform measured with respect to the XYZ coordinate system.

(c) Assuming that the magnitude of the forces acting along the legs are measured as:

Leg	Force (N)
$1 - 4$	20
$1 - 5$	12
$2 - 5$	20
$2 - 6$	14
$3 - 6$	17
$3 - 4$	24

where a positive force value indicates that the leg is in compression, determine the external wrench, measured in the XYZ coordinate system, that is acting on top platform if the system is in static equilibrium at this instant.

(d) Assuming now that an external wrench $\hat{w}$ is acting on the top platform and its coordinates as measured in the XYZ system are $\hat{w} = [5, 8, 12; 20, -15, 20]$, where the first three components have units of N, and the last three have units

of Nm, determine the forces in the six legs such that the platform is in static
equilibrium.

5. Two wrenches $\hat{w}_1$ and $\hat{w}_2$ whose coordinates are $[5, -4, 2; 30, -15, 20]$ and
$[-6, 8, -2; -20, 10, 10]$, respectively, are acting on a rigid body. The first three
components of the wrenches have units of N, and the last three have units of Nm.
   (a) Determine the coordinates of a third wrench that will hold the rigid body in
   static equilibrium.
   (b) Determine the magnitude of the force along the third wrench, its line of
   action, and its pitch.

6. A cylindroid is generated from the two screws $\hat{s}_1 = [1, 0, 0; 3, 6, -12]$ and
$\hat{s}_2 = [0, \frac{\sqrt{2}}{2}, -\frac{\sqrt{2}}{2}; 20, -15, 20]$. The coordinates of the screws are written with
respect to the XYZ coordinate system; the first three components of each screw
are dimensionless, and the last three components have units of meters.
   (a) Determine the coordinates of the point of intersection of the mutually
   perpendicular line with the first screw.
   (b) Determine the coordinates of the point of intersection of the mutually
   perpendicular line with the second screw.
   (c) Determine the Plücker coordinates of the line that is mutually perpendicular
   to the lines of action of the two screws.
   (d) Write the coordinates of the two screws in terms of a new xyz coordinate
   system whose origin is located at the intersection of the first screw and the
   mutually perpendicular line, whose z axis is along the mutually perpendicular
   line, and whose x axis is along the first screw.
   (e) A cylindroid is generated from all the linear combinations of the two
   screws as

$$\hat{w} = f_1 \hat{s}_1 + f_2 \hat{s}_2.$$

   (i) Determine the minimum and maximum pitch values for the cylindroid.
   (ii) Determine the ratio of $\frac{f_1}{f_2}$, such that the resultant wrench has its
   minimum possible pitch.
   (iii) Determine the ratio of $\frac{f_1}{f_2}$, such that the resultant wrench has its
   maximum possible pitch.
   (iv) Determine the length of the cylindroid.
   (v) Determine the coordinates of the center point of the cylindroid in the
   original XYZ coordinate system and in the new xyz coordinate system.

4   Velocity Analysis

A body is said to receive a *twist* about a screw
when it is rotated uniformly about the screw,
while it is translated uniformly parallel to the
screw, through a distance equal to the product of
the pitch and the circular measure of the angle of
rotation.

Ball (1900)

4.1   Introduction

Suppose that a rigid body B is in motion relative to a rigid body A. At some instant in time, it would be possible to measure the velocity of any or all points in body B in terms of a coordinate system fixed in body A. Recording the velocities of all points in body B, however, is not a practical means of describing the instantaneous velocity state of body B. One objective of this chapter is to select a minimum number of parameters that will completely describe this velocity state. From these parameters it will be possible to determine the velocity of any point in body B.

Since body B is a rigid body, it must physically be the case that the instantaneous motion of body B with respect to body A can always be described by the combination of a translation in some direction and a rotation about some instantaneously fixed line in space. It will be shown in this chapter that the velocity state of body B can indeed be defined in terms of these concepts. To obtain a final expression for the velocity state of body B, the chapter will proceed by first determining the time derivative of an arbitrary vector that is in motion relative to its reference coordinate system.

4.2   Time Derivative of a Vector

A vector is defined as a physical quantity with both magnitude and direction, and vectors are often used to represent forces, moments, linear and angular velocities, and linear and angular accelerations. Here, the focus will be on obtaining the time derivative of vectors that represent the relative position of two points. The vector under consideration will originate at a point A and terminate at a point B, and it is often said that point A is the "tail" of the vector and point B is the "head" of the vector. Five cases will be considered where in each case the derivative of the vector will be obtained with respect to a reference frame attached to body 0.

4.2.1   Case 1: Points *A* and *B* Embedded in Body 0

Figure 4.1 shows the simple case where the head and tail of vector α (points B and A, respectively) are both embedded in body 0. The time rate of change of vector α measured with respect to body 0 is obviously the zero vector and thus

$$\frac{{}^0d}{dt}\alpha = \mathbf{0}, \tag{4.1}$$

where the superscript 0 indicates that the time derivative is evaluated with respect to a reference system embedded in body 0.

4.2.2   Case 2: Point *A* Embedded in Body 0 and Point *B* Embedded in Body 1

The second case is shown in Figure 4.2. Here the "tail" of vector β (point A) is fixed in body 0 while the "head" of the vector (point B) is embedded in body 1. The vector β may be written as

$$^0\beta = {}^0p^1_B - {}^0p^0_A, \tag{4.2}$$

where the notation $^i\rho$ indicates that the vector ρ is being measured with respect to the reference frame embedded in body i and the notation $^ip^j_k$ represents the position of point k, which is embedded in body j measured in terms of a reference coordinate system embedded in body i. The time rate of change of β measured with respect to body 0 can be expressed as

$$\frac{{}^0d}{dt}\beta = \frac{{}^0d}{dt}{}^0p^1_B - \frac{{}^0d}{dt}{}^0p^0_A. \tag{4.3}$$

The second term on the right side of this equation is obviously 0 (see case 1). The first term represents the time rate of change of a vector from the origin of the reference

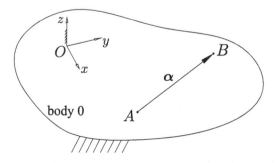

Figure 4.1 Definition of vector α

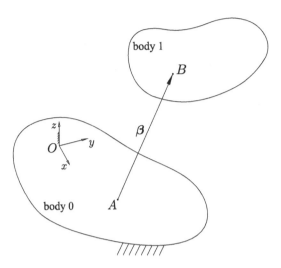

Figure 4.2 Definition of vector β

frame to the moving point B. This is obviously the velocity of point B measured with respect to the reference frame embedded in body 0 and thus

$$\frac{^0d}{dt}\beta = {}^0v_B^1, \tag{4.4}$$

where the notation ${}^iv_k^j$ represents the velocity of point k, which is embedded in body j measured with respect to a reference frame embedded in body i.

4.2.3   Case 3: Points *A* and *B* Embedded in Body 1

Figure 4.3 shows the case where the vector γ is defined as having its "tail" at point A and its "head" at point B where both points are embedded in body 1. It is desired to obtain the time derivative of the vector γ with respect to a reference frame embedded in body 0. The vector γ may be written in terms of the reference system attached to body 0 as

$$^0\gamma = {}^0p_B^1 - {}^0p_A^1. \tag{4.5}$$

Thus

$$\frac{^0d}{dt}\gamma = \frac{^0d}{dt}\,{}^0p_B^1 - \frac{^0d}{dt}\,{}^0p_A^1. \tag{4.6}$$

Based on the result of case 2, this may be written as

$$\frac{^0d}{dt}\gamma = {}^0v_B^1 - {}^0v_A^1. \tag{4.7}$$

Now consider that for a general case any vector can change with respect to time in two ways, i.e., the vector may change in length (magnitude) and it may change orientation. In this case, since points A and B are embedded in the same body, the

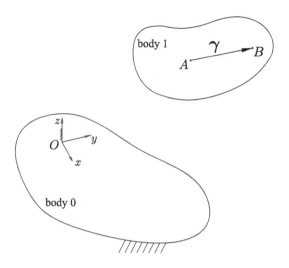

Figure 4.3 Definition of vector γ

vector γ will not change in length when observed from the reference frame embedded in body 0. As previously mentioned, the instantaneous motion of body 1 with respect to body 0 can always be described by the combination of a translation in some direction and a rotation about some instantaneously fixed line in space. For this case, it is obvious that the translation of body 1 with respect to body 0 will not cause the vector γ to change in direction and, thus, this translation term can be ignored when calculating $\frac{^0d}{dt}\gamma$. Figure 4.4 shows the particular axis that body 1 is at this instant rotating about relative to body 0. The axis passes through point M, and the direction of the axis is parallel to the unit vector $^0S^1$. The magnitude of the angular velocity is $_0\omega_1$. The angular velocity vector will be defined as $^0\omega^1$, where

$$^0\omega^1 = {_0}\omega_1\, ^0S^1. \tag{4.8}$$

The velocity of point A measured with respect to body 0 can be calculated as the cross product of the angular velocity vector and a vector from any point on the rotation axis to point A. Thus, the velocity of point A may be written as

$$^0v_A^1 = {^0\omega^1} \times \left({^0p_A^1} - {^0p_M^1}\right). \tag{4.9}$$

Similarly, the velocity of point B measured with respect to body 0 can be calculated as

$$^0v_B^1 = {^0\omega^1} \times \left({^0p_B^1} - {^0p_M^1}\right). \tag{4.10}$$

Substituting (4.9) and (4.10) into (4.7) yields

$$\frac{^0d}{dt}\gamma = {^0\omega^1} \times \left({^0p_B^1} - {^0p_M^1}\right) - {^0\omega^1} \times \left({^0p_A^1} - {^0p_M^1}\right). \tag{4.11}$$

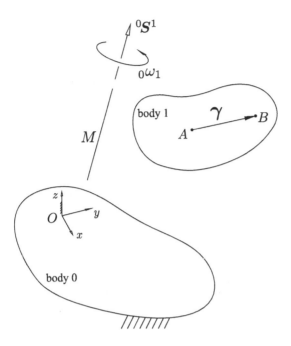

Figure 4.4 Instantaneous axis of rotation

Regrouping this result gives

$$\frac{^0d}{dt}\gamma = {}^0\boldsymbol{\omega}^1 \times \left({}^0\boldsymbol{p}_B^1 - {}^0\boldsymbol{p}_A^1\right) = {}^0\boldsymbol{\omega}^1 \times {}^0\gamma. \tag{4.12}$$

4.2.4   Case 4: Point *A* Embedded in Body 1 and Point *B* Embedded in Body 2

Figure 4.5 shows the case where the "tail" of vector ζ, point A, is embedded in body 1 and the "head" of the vector, point B, is embedded in body 2. It is desired to obtain the time derivative of the vector ζ with respect to a reference frame embedded in body 0. The vector ζ measured with respect to the reference frame may be written as

$$^0\zeta = {}^0\boldsymbol{p}_B^2 - {}^0\boldsymbol{p}_A^1. \tag{4.13}$$

The vector $^0\boldsymbol{p}_B^2$ can be written as the sum of a vector from the origin of the reference system to a point embedded in body 1 that is coincident with point B, i.e., $^0\boldsymbol{p}_B^1$, plus a vector from this point to point B that is embedded in body 2. Thus

$$^0\boldsymbol{p}_B^2 = {}^0\boldsymbol{p}_B^1 + \left({}^0\boldsymbol{p}_B^2 - {}^0\boldsymbol{p}_B^1\right). \tag{4.14}$$

Equation (4.13) can thus be written as

$$^0\zeta = \left({}^0\boldsymbol{p}_B^1 - {}^0\boldsymbol{p}_A^1\right) + \left({}^0\boldsymbol{p}_B^2 - {}^0\boldsymbol{p}_B^1\right), \tag{4.15}$$

and the time derivative of the vector ζ can be written as

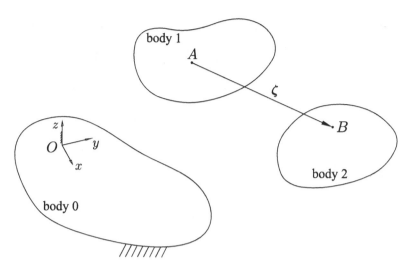

Figure 4.5 Definition of vector ζ

$$\frac{{}^0d}{dt}\zeta = \frac{{}^0d}{dt}\left({}^0p_B^1 - {}^0p_A^1\right) + \frac{{}^0d}{dt}\left({}^0p_B^2 - {}^0p_B^1\right).$$ (4.16)

The first term is the time derivative of a vector whose "tail" and "head" are both embedded in body 1. From (4.12), this derivative can be written as

$$\frac{{}^0d}{dt}\left({}^0p_B^1 - {}^0p_A^1\right) = {}^0\omega^1 \times \left({}^0p_B^1 - {}^0p_A^1\right).$$ (4.17)

At the instant under consideration, since the coordinates of point B embedded in body 1 and point B embedded in body 2 are identical, the coordinates of ζ measured with respect to the reference frame embedded in body 0 will equal $\left({}^0p_B^1 - {}^0p_A^1\right)$, and (4.17) may be written as

$$\frac{{}^0d}{dt}\left({}^0p_B^1 - {}^0p_A^1\right) = {}^0\omega^1 \times {}^0\zeta.$$ (4.18)

The second term on the right side of (4.16) is yet to be determined. This term may be written as

$$\frac{{}^0d}{dt}\left({}^0p_B^2 - {}^0p_B^1\right) = \frac{{}^0d}{dt}\,{}^0p_B^2 - \frac{{}^0d}{dt}\,{}^0p_B^1.$$ (4.19)

From case 2, this equation may be written as

$$\frac{{}^0d}{dt}\left({}^0p_B^2 - {}^0p_B^1\right) = {}^0v_B^2 - {}^0v_B^1,$$ (4.20)

where the right side terms represent the difference between the velocities of point B in body 2 and point B in body 1, both measured with respect to body 0. This can also be physically interpreted as the velocity of point B in body 2 seen with respect to an observer in body 1. The reference frame attached to body 1 to be used here will be coincident and aligned with the reference frame attached to body 0, i.e., at this instant the two reference frames have the same origin point and their corresponding coordinate axes are parallel. For this case then, the velocity of point B in body 2 seen with respect to the reference frame attached to body 1 can be written as

$$^1v_B^2 = {}^0v_B^2 - {}^0v_B^1 = \frac{{}^1d}{dt}{}^1p_B^2.$$

(4.21)

The vector $^1p_B^2$ can be written as the sum of a vector from the origin of the reference frame in body 1 to point A in body 1 plus the vector from point A in body 1 to point B in body 2. Thus

$$^1p_B^2 = {}^1p_A^1 + {}^1\zeta$$

(4.22)

and

$$\frac{{}^1d}{dt}{}^1p_B^2 = \frac{{}^1d}{dt}{}^1\zeta.$$

(4.23)

Substituting (4.23) into (4.21) and then this result into (4.20) yields

$$\frac{{}^0d}{dt}({}^0p_B^2 - {}^0p_B^1) = \frac{{}^1d}{dt}{}^1\zeta.$$

(4.24)

Substituting (4.24) and (4.18) into (4.16) gives

$$\frac{{}^0d}{dt}\zeta = {}^0\omega^1 \times {}^0\zeta + \frac{{}^1d}{dt}{}^1\zeta.$$

(4.25)

Since the reference coordinate systems attached to body 0 and body 1 are at this instant coincident and aligned, it will be the case that $^0\zeta = {}^1\zeta$ and often the terms $^0\zeta$ and $^1\zeta$ in the equation are written without the preceding superscripts. For a general case where the "tail" of the vector ζ, point A, lies in body i and the "head" of the vector ζ, point B, lies in body j, the derivative of ζ with respect to a coordinate system embedded in body 0 may be written as

$$\frac{{}^0d}{dt}\zeta = {}^0\omega^i \times {}^0\zeta + \frac{{}^id}{dt}{}^i\zeta.$$

(4.26)

4.2.5   Case 5: Derivatives of a Vector with Respect to Two Coordinate Systems

Figure 4.6 shows two bodies and an arbitrary vector ξ. Body 1 is moving with respect to body 0, and at the instant shown the angular velocity of body 1 with respect to body 0 is given by $^0\omega^1$. Without loss of generality, it is assumed here that the reference coordinate system attached to body 1 is at this instant coincident and aligned with that of body 0. It is desired to relate the time derivatives of vector ξ as measured with respect to body 0 and body 1.

Now, let i_1, j_1, and k_1 be a set of mutually perpendicular vectors embedded in body 1. The vector ξ may be written in terms of these vectors as

$$\xi = a\,i_1 + b\,j_1 + c\,k_1.$$

(4.27)

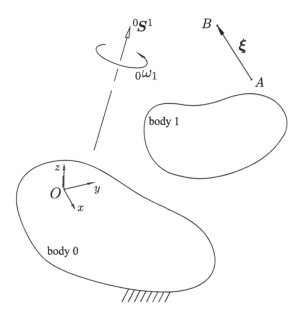

Figure 4.6 Depiction of vector ξ

Taking the derivative of this vector with respect to body 0 gives

$$\frac{^0d}{dt}\xi = \left(\frac{^0d}{dt}a\,i_1 + \frac{^0d}{dt}b\,j_1 + \frac{^0d}{dt}c\,k_1\right) + \left(a\,\frac{^0d}{dt}i_1 + b\,\frac{^0d}{dt}j_1 + c\,\frac{^0d}{dt}k_1\right). \quad (4.28)$$

The first term in parentheses contains the time rate of change of the scalars a, b, and c as measured with respect to body 0. The time rate of change of these scalars is independent of the choice of coordinate system, and thus

$$\frac{^0d}{dt}a\,i_1 + \frac{^0d}{dt}b\,j_1 + \frac{^0d}{dt}c\,k_1 = \frac{^1d}{dt}a\,i_1 + \frac{^1d}{dt}b\,j_1 + \frac{^1d}{dt}c\,k_1. \quad (4.29)$$

It is recognized that the right side of (4.29) is the time derivative of the vector ξ measured with respect to the reference system embedded in body 1, and thus

$$\frac{^0d}{dt}a\,i_1 + \frac{^0d}{dt}b\,j_1 + \frac{^0d}{dt}c\,k_1 = \frac{^1d}{dt}\xi. \quad (4.30)$$

The terms $\frac{^0d}{dt}i_1$, $\frac{^0d}{dt}j_1$, and $\frac{^0d}{dt}k_1$ in the second term in parentheses in (4.28) represent the time derivative of vectors that are embedded in body 1 with respect to the reference frame embedded in body 0. From (4.12) of case 3, it may be written that

$$a\,\frac{^0d}{dt}i_1 + b\,\frac{^0d}{dt}j_1 + c\,\frac{^0d}{dt}k_1 = a\left(^0\omega^1 \times i_1\right) + b\left(^0\omega^1 \times j_1\right) + c\left(^0\omega^1 \times k_1\right). \quad (4.31)$$

This may be regrouped as

$$a\,\frac{^0d}{dt}i_1 + b\,\frac{^0d}{dt}j_1 + c\,\frac{^0d}{dt}k_1 = {}^0\omega^1 \times \left(a\,i_1 + b\,j_1 + c\,k_1\right) \quad (4.32)$$

and upon substituting (4.27) yields

$$a\,\frac{^0d}{dt}i_1 + b\,\frac{^0d}{dt}j_1 + c\,\frac{^0d}{dt}k_1 = {}^0\omega^1 \times \xi.\tag{4.33}$$

Substituting (4.30) and (4.33) into (4.28) gives

$$\frac{^0d}{dt}\xi = \frac{^1d}{dt}\xi + {}^0\omega^1 \times \xi.\tag{4.34}$$

This is a general result that relates the time derivatives of a vector that are taken with respect to two bodies that are in relative motion.

4.3   Definition of Velocity State

The velocity state of a rigid body is to be defined by a set of parameters from which the velocity of any point in the rigid body may be determined. It is proposed here that the instantaneous velocity state of body 1 measured with respect to body 0 be defined by the angular velocity vector ${}^0\omega^1$ and the position and linear velocity of any one point (called point R) in body 1, i.e., ${}^0p_R^1$ and ${}^0v_R^1$ (see Figure 4.7). From these parameters, it will be shown that the velocity of any other point in body 1 may be determined with respect to body 0.

The vector r, as shown in Figure 4.8, is drawn from point O that is fixed in body 0 to point R that is fixed in body 1. The vector p is drawn from point O in body 0 to point P in body 1. The vector $r_{R\rightarrow P}$ is drawn from point R in body 1 to point P in body 1 and is thus fixed in body 1.

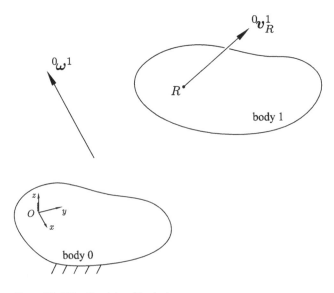

Figure 4.7 Velocity state of body 1

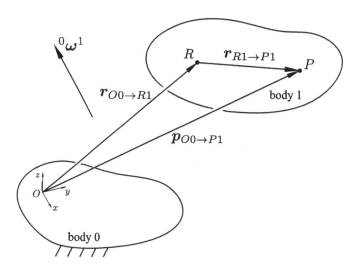

Figure 4.8 Determination of velocity of point P

From Figure 4.8, it is apparent that

$$\boldsymbol{p}_{O0 \to P1} = \boldsymbol{r}_{O0 \to R1} + \boldsymbol{r}_{R1 \to P1}. \qquad (4.35)$$

Taking a time derivative of this equation with respect to body 0 gives

$$\frac{^0 d}{dt}\boldsymbol{p}_{O0 \to P1} = \frac{^0 d}{dt}\boldsymbol{r}_{O0 \to R1} + \frac{^0 d}{dt}\boldsymbol{r}_{R1 \to P1}. \qquad (4.36)$$

Vectors $\boldsymbol{r}_{O0 \to R1}$ and $\boldsymbol{p}_{O0 \to P1}$ both have their "tails" embedded in body 0 and their "heads" embedded in body 1. Thus, from case 2 of Section 4.2

$$\frac{^0 d}{dt}\boldsymbol{r}_{O0 \to R1} = {}^0\boldsymbol{v}_R^1, \qquad (4.37)$$

$$\frac{^0 d}{dt}\boldsymbol{p}_{O0 \to R1} = {}^0\boldsymbol{v}_P^1. \qquad (4.38)$$

Since the vector $\boldsymbol{r}_{R1 \to P1}$ is embedded in body 1, equation (4.12) of case 3 of Section 4.2 can be used to evaluate the time derivative of this vector with respect to a reference frame in body 0 as

$$\frac{^0 d}{dt}\boldsymbol{r}_{R1 \to P1} = {}^0\boldsymbol{\omega}^1 \times {}^0\boldsymbol{r}_{R1 \to P1}. \qquad (4.39)$$

Substituting (4.37), (4.38), and (4.39) into (4.36) gives

$$^0\boldsymbol{v}_P^1 = {}^0\boldsymbol{v}_R^1 + {}^0\boldsymbol{\omega}^1 \times {}^0\boldsymbol{r}_{R1 \to P1}. \qquad (4.40)$$

Equation (4.40) may be used to determine the velocity of any point in body 1 with respect to body 0 from the velocity state parameters ${}^0\boldsymbol{\omega}^1$ and ${}^0\boldsymbol{v}_R^1$. It is often convenient to have the reference point R be the point in body 1 that is instantaneously coincident with point O, the origin of the reference frame in body 0. For this case, equation (4.40) becomes

$$^0v_P^1 = {}^0v_O^1 + {}^0\omega^1 \times {}^0r_{O1 \to P1}.$$ (4.41)

Here, the notation $^0r_{O1 \to P1}$ is used to emphasize that this vector originates at point O in body 1 and ends at point P, which is also in body 1. Although this appears immaterial, since the vector from point O in body 0 to point P in body 1 is at this instant the same, it will be important in Chapter 7 when the derivative of equation (4.41) will be taken with respect to body 0 as part of an acceleration analysis. Lastly, (4.41) is often written as

$$^0v_P^1 = {}^0v_O^1 + {}^0\omega^1 \times p,$$ (4.42)

where p represents the coordinates of point P as measured in the reference frame attached to body 0. The notation $\{{}^0\omega^1; {}^0v_O^1\}$ will be used to represent the velocity state. The semicolon is used to emphasize that the units of the two terms are different, i.e., rad/sec and length/sec. It should be apparent that the velocity state as specified in (4.42) by the terms $^0\omega^1$ and $^0v_O^1$ can be interpreted as body 1 rotating about a line parallel to $^0\omega^1$ that passes through the origin with an angular velocity of $_0\omega_1 = |{}^0\omega^1|$ while simultaneously translating with a velocity of $^0v_O^1$ with respect to body 0.

The states of rotating about an axis that passes through the origin and translating will now be considered separately. The velocity state of a body that is rotating about an axis parallel to $^0\omega^1$ that passes through the origin can be written as $\{{}^0\omega^1; 0\}$. This may also be written as

$$\{{}^0\omega^1; 0\} = {}_0\omega_1\{{}^0S^1; 0\},$$ (4.43)

where $^0S^1$ is a unit vector defined as

$$^0S^1 = \frac{{}^0\omega^1}{{}_0\omega_1}.$$ (4.44)

The term $\{{}^0S^1; 0\}$ represents the Plücker coordinates of the line along the axis of rotation through the origin.

The velocity state of a body that is in pure translation with a velocity of $^0v^1$ can be written as $\{0; {}^0v^1\}$. This can readily be inferred, since in this case the body has no angular velocity, i.e., $_0\omega_1 = 0$, and the velocity of the point in the body coincident with the origin (and the velocity of all points in the body) is equal to $^0v^1$. The translation with a velocity of $^0v^1 = {}_0v_1 S_v$, where S_v is a unit vector, can be considered as being equivalent to a rotation with an angular velocity $\delta^0\omega^1$ of infinitesimal magnitude ($|\delta^0\omega^1| \to 0$) about a line that is infinitely distant. The velocity of any point in the body can be calculated as $^0v^1 = \delta^0\omega^1 \times r$, where r is a vector from a point on the axis of rotation to a point on the moving body such that $|r| = \infty$. The line of action of $\delta^0\omega^1$ is at infinity, and the Plücker coordinates of this line may be written as $\{0; S_v\}$ (see depiction in Figure 4.9). Thus, the velocity state of a body in pure translation can be written as the product of the magnitude of the translational velocity times the Plücker line coordinates of a line at infinity as

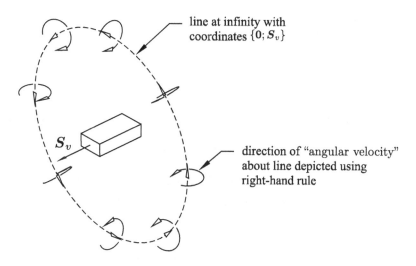

line at infinity with
coordinates $\{0; S_v\}$

S_v

direction of "angular velocity"
about line depicted using
right-hand rule

Figure 4.9 Depiction of line at infinity

$$\{0;\ {}^0v^1\} = {}_0v_1\{0;\ S_v\}. \tag{4.45}$$

Finally, the general velocity state $\{{}^0\boldsymbol{\omega}^1;\ {}^0v_O^1\}$ can be thought of as a rotation about
a line passing through the origin, i.e., $\{{}^0S^1; \mathbf{0}\}$, where ${}^0S^1$ is a unit vector, with a
rotational velocity of ${}_0\omega_1$ together with a translation of ${}^0v_O^1 = {}_0v_{O1}\, S_{Ov}$, where S_{Ov}
is a dimensionless unit vector in the direction of the translation and ${}_0v_{O1}$ is the linear
velocity magnitude. This can be written as

$$\{{}^0\boldsymbol{\omega}^1;\ {}^0v_O^1\} = \{{}^0\boldsymbol{\omega}^1; \mathbf{0}\} + \{\mathbf{0};\ {}^0v_O^1\}$$
$$= {}_0\omega_1\, \{{}^0S^1; \mathbf{0}\} + {}_0v_{O1}\, \{\mathbf{0};\ S_{Ov}\}, \tag{4.46}$$

4.4   Screw Interpretation of Velocity State

In the previous section, it was shown that the velocity state of body 1 can be described
by the parameters ${}^0\boldsymbol{\omega}^1$ and ${}^0v_O^1$. The velocity of the point in body 1 that is coincident
with the origin point O can be separated into two components, one parallel to ${}^0\boldsymbol{\omega}^1$,
i.e., ${}^0v_{Oa}^1$, and one perpendicular, i.e., ${}^0v_{Ot}^1$, as

$$ {}^0v_O^1 = {}^0v_{Ot}^1 + {}^0v_{Oa}^1, \tag{4.47}$$

where

$$ {}^0v_{Oa}^1 = \left(\frac{{}^0\boldsymbol{\omega}^1}{{}_0\omega_1} \cdot {}^0v_O^1 \right) \frac{{}^0\boldsymbol{\omega}^1}{{}_0\omega_1} \tag{4.48}$$

and

$$ {}^0v_{Ot}^1 = {}^0v_O^1 - {}^0v_{Oa}^1. \tag{4.49}$$

The velocity state may now be written as

$$\{{}^0\omega^1; {}^0v_O^1\} = \{{}^0\omega^1; {}^0v_{Ot}^1\} + \{\mathbf{0}; {}^0v_{Oa}^1\}. \tag{4.50}$$

Since ${}^0\omega^1$ is perpendicular to ${}^0v_{Ot}^1$, the first term on the right side of (4.50) can be interpreted as a rotation of magnitude ${}_0\omega_1$ about a line whose Plücker coordinates are given by $\left\{\frac{{}^0\omega^1}{{}_0\omega_1}; \frac{{}^0v_{Ot}^1}{{}_0\omega_1}\right\}$. This is confirmed by calculating the velocity of the point in body 1 coincident with the origin point if body 1 were indeed rotating about this line with a magnitude of ${}_0\omega_1$.

The second term of (4.50) can be interpreted as a translation in a direction parallel to ${}^0\omega^1$. Since ${}^0\omega^1$ and ${}^0v_{Oa}^1$ are parallel, they can be related as

$$ {}^0v_{Oa}^1 = h\,{}^0\omega^1, \tag{4.51}$$

where h will be referred to as the pitch, which is clearly the ratio of the magnitude of ${}^0v_{Oa}^1$ to the magnitude of ${}^0\omega^1$. Performing a scalar product of both sides of (4.51) with ${}^0\omega^1$ gives

$$ {}^0\omega^1 \cdot {}^0v_{Oa}^1 = h\left({}^0\omega^1 \cdot {}^0\omega^1\right). \tag{4.52}$$

Solving for the pitch h gives

$$ h = \frac{{}^0\omega^1 \cdot {}^0v_{Oa}^1}{{}^0\omega^1 \cdot {}^0\omega^1}. \tag{4.53}$$

Substituting (4.47) into (4.53) gives

$$ h = \frac{{}^0\omega^1 \cdot \left({}^0v_O^1 - {}^0v_{Ot}^1\right)}{{}^0\omega^1 \cdot {}^0\omega^1}, \tag{4.54}$$

which may be simplified since ${}^0\omega^1$ and ${}^0v_{Ot}^1$ are perpendicular as

$$ h = \frac{{}^0\omega^1 \cdot {}^0v_O^1}{{}^0\omega^1 \cdot {}^0\omega^1}. \tag{4.55}$$

Substituting (4.49) and (4.51) into (4.50) yields

$$\{{}^0\omega^1; {}^0v_O^1\} = \{{}^0\omega^1; {}^0v_O^1 - h\,{}^0\omega^1\} + \{\mathbf{0}; h\,{}^0\omega^1\}. \tag{4.56}$$

This equation represents the screw interpretation of the velocity state. The velocity state can thus be interpreted as the combination of a rotation of magnitude ${}_0\omega_1$ about the line whose Plücker coordinates are defined by

$$\{S; S_{OL}\} = \left\{\frac{{}^0\omega^1}{{}_0\omega_1}; \frac{{}^0v_O^1 - h\,{}^0\omega^1}{{}_0\omega_1}\right\} = \frac{1}{{}_0\omega_1}\left\{{}^0\omega^1; {}^0v_O^1 - h\,{}^0\omega^1\right\} \tag{4.57}$$

together with a translation of magnitude $h\,{}_0\omega_1$ about a line at infinity whose Plücker line coordinates are defined as $\left\{\mathbf{0}; \frac{{}^0\omega^1}{{}_0\omega_1}\right\}$. Alternatively, the velocity state can be considered as motion about the screw $\left\{\frac{{}^0\omega^1}{{}_0\omega_1}; \frac{{}^0v_O^1}{{}_0\omega_1}\right\}$ with a magnitude of ${}_0\omega_1$.

From (4.57), it is apparent that the moment of the line of action of the screw can be expressed as

$$S_{OL} = \frac{{}^0 v_O^1 - h^0 \omega^1}{{}_0 \omega_1}.$$ (4.58)

Rearranging this equation yields

$${}^0 v_O^1 = {}_0 \omega_1 S_{OL} + h^0 \omega^1 = {}_0 \omega_1 (S_{OL} + h\, S).$$ (4.59)

This equation is useful for the case where the line of action of the screw is given as $\{S; S_{OL}\}$ together with the pitch, h, and angular velocity ${}_0\omega_1$, and it is desired to determine the velocity state components $\{{}^0\omega^1; {}^0 v_O^1\}$.

In Chapter 3, a screw was defined as a line with a pitch. Multiplying a screw by a force magnitude yielded a wrench. Here, multiplying a screw by an angular velocity magnitude yields a twist. A screw is represented by the symbol $\$$ or $\hat{s}$, a wrench by the symbol $\hat{w}$, and a twist by the symbol $\hat{T}$.

4.4.1   Example Problem 1

A velocity state is given as ${}^0\omega^1 = [0,0,2]^T$ rad/sec and ${}^0 v_O^1 = [2,-4,2]^T$ m/sec. Determine the pitch of the screw and the point in the XY plane that lies on the axis of rotation.

The pitch of the screw is calculated from (4.55) as

$$h = \frac{{}^0\omega^1 \cdot {}^0 v_O^1}{{}^0\omega^1 \cdot {}^0\omega^1} = \frac{4}{4} = 1 \text{ m.}$$

The Plücker coordinates of the line of action of the screw can be calculated from (4.57) as

$$\{S; S_{OL}\} = \frac{1}{{}_0\omega_1} \left\{ {}^0\omega^1; {}^0 v_O^1 - h^0\omega^1 \right\},$$

where ${}_0\omega_1 = |{}^0\omega^1| = 2$ rad/sec. Thus

$$\{S; S_{OL}\} = \{0,0,1; 1,-2,0\},$$

where the first three components are dimensionless and the last three have units of meters. A point on this line can be determined from

$$r \times S = S_{OL}.$$

Writing $r = r_x i + r_y j + r_z k$ gives

$$(r_x i + r_y j + r_z k) \times k = i - 2j.$$

Expanding this equation yields

$$r_x = 2, \quad r_y = 1.$$

Thus, points on the axis of rotation have coordinates $(2, 1, z)$ and the axis of rotation intersects the XY plane at $(2, 1, 0)$.

Figure 4.10 shows the velocity vector field for several points in body 1 where (4.42) was used to calculate the velocity for each of the points in the figure. It is important to recognize that every point in the figure also has a velocity of 2 m/sec along the Z direction, out of the page.

One interpretation of the given velocity state would be body 1 simultaneously rotating about a line parallel to $^0\omega^1$, which passes through point O, and translating with a velocity of $^0v_O^1$. Figure 4.11(a) shows the velocity field due to the rotation about point O, while Figure 4.11(b) shows the translational velocity field. Figure 4.11(c) shows the total velocity field, which is the same as that shown in Figure 4.10. In Figure 4.12, the three figures shown in Figure 4.11 are superimposed so that it can be

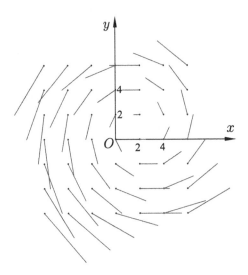

Figure 4.10 Velocity of points in body 1

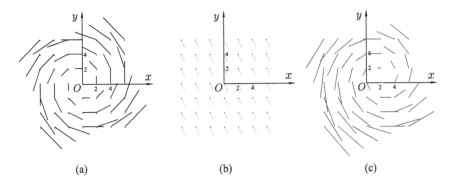

(a)   (b)   (c)

Figure 4.11 Velocity of points in body 1: (a) rotational velocity field, (b) translational velocity field, and (c) total velocity field

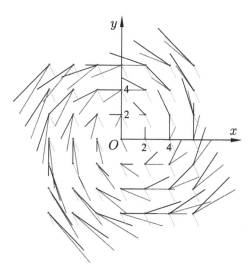

Figure 4.12 Velocity of points in body 1

more readily seen at each point that the total velocity field is indeed equal to the sum of the rotational and translational velocity fields.

Figure 4.10 also shows the screw interpretation of the velocity state. It is apparent in the figure that there is a rotation about the point $(2, 1, 0)$, which is a point on the line of action of the twist. Coming out of the page, and parallel to the axis of rotation, is the translational velocity component of 2 m/sec.

4.4.2   Example Problem 2

The angular velocity of body 1 measured with respect to body 0 is given as $^0\omega^1 = [2, 4, 6]^T$ rad/sec. The coordinates of point A in body 1 are given in terms of the reference coordinate system in body 0 as $^0P_A^1 = [4, 4, -2]^T$ meters, and the velocity of this point is given as $^0v_A^1 = [6, -8, 8]^T$ m/sec. Determine the velocity state parameter $^0v_O^1$.

From (4.42), a relationship between the given velocity $^0v_A^1$ and $^0v_O^1$ can be written as

$$^0v_A^1 = {}^0v_O^1 + {}^0\omega^1 \times {}^0P_A^1.$$

Solving this equation for $^0v_O^1$ gives

$$^0v_O^1 = {}^0v_A^1 - {}^0\omega^1 \times {}^0P_A^1.$$

Substituting the given values gives

$$^0v_O^1 = [38, -36, 16]^T \text{ m/sec.}$$

4.4.3   Example Problem 3

The velocity state of body 1 with respect to body 0 is given by the parameters
$^0\omega^1 = [-1, 3, 2]^T$ rad/sec and $^0v_O^1 = [4, -2, 8]^T$ m/sec. Determine the velocity of
point A in body 1 where its coordinates are given as $^0P_A^1 = [8, 2, -4]^T$ m. Determine
the pitch of the screw and the Plücker line coordinates of the screw line of action.
Determine a point on the line of action of the screw.

The velocity of point A in body 1 can be determined from (4.42) as

$$^0v_A^1 = {}^0v_O^1 + {}^0\omega^1 \times {}^0P_A^1$$

$$= \begin{bmatrix} 4 \\ -2 \\ 8 \end{bmatrix} + \begin{bmatrix} -1 \\ 3 \\ 2 \end{bmatrix} \times \begin{bmatrix} 8 \\ 2 \\ -4 \end{bmatrix} = \begin{bmatrix} -12 \\ 10 \\ -18 \end{bmatrix} \text{ m/sec.}$$

The pitch of the screw can be determined from (4.55) to give the result

$$h = \frac{{}^0\omega^1 \cdot {}^0v_O^1}{{}^0\omega^1 \cdot {}^0\omega^1}$$

$$= \frac{[-1, 3, 2]^T \cdot [4, -2, 8]^T}{[-1, 3, 2]^T \cdot [-1, 3, 2]^T} = \frac{6}{14} = 0.429 \text{ m.}$$

The Plücker coordinates of the line of axis of the screw can be determined from
(4.57) to be

$$\{S; S_{OL}\} = \frac{1}{{}^0\omega_1} \left\{ {}^0\omega^1; {}^0v_O^1 - h\,{}^0\omega^1 \right\}$$

$$= \frac{1}{\sqrt{14}} \left\{ [-1, 3, 2]^T; [4, -2, 8]^T - \frac{6}{14}[-1, 3, 2]^T \right\}$$

$$= \{-0.267, 0.802, 0.535 ; 1.184, -0.878, 1.909\},$$

where the first three components of the line are dimensionless while the last three
components have dimensions of m.

Equation (1.59) may used to determine the point on the line of action of the screw
that is closest to the origin. Thus

$$p = \frac{S \times S_{OL}}{S \cdot S} = [2, 1.143, -0.714]^T \text{ m.}$$

4.4.4   Example Problem 4

Body 1 is constrained to move about a screw with respect to body 0. The line of
action of the screw is given by the unit Plücker line coordinates $\{S; S_{OL}\} = \{\sqrt{2}/2,$
$-\sqrt{2}/2, 0 ; 5, 5, -2\}$, where the first three parameters are dimensionless while the last
three parameters have dimensions of m. The pitch of the screw is equal to $h = 2.2$ m.
The magnitude of the angular velocity of body 1 about the screw is equal to $_0\omega_1 = 2.5$
rad/sec. Determine the velocity state of body 1 with respect to body 0.

The angular velocity of body 1 with respect to body 0 may be determined as the product of the magnitude of the rotational velocity times the unit directional vector of the line of action of the screw. Thus

$$ {}^0\boldsymbol{\omega}^1 = {}_0\omega_1\,\boldsymbol{S} = 2.5 \begin{bmatrix} 0.707 \\ -0.707 \\ 0 \end{bmatrix} = \begin{bmatrix} 1.768 \\ -1.768 \\ 0 \end{bmatrix} \text{ rad/sec.} $$

From (4.59), ${}^0\boldsymbol{v}_O^1$ may be determined as

$$ {}^0\boldsymbol{v}_O^1 = {}_0\omega_1\,(\boldsymbol{S}_{OL} + h\,\boldsymbol{S}) $$

$$ = 2.5 \left(\begin{bmatrix} 5 \\ 5 \\ -2 \end{bmatrix} + 2.2 \begin{bmatrix} 0.707 \\ -0.707 \\ 0 \end{bmatrix} \right) = \begin{bmatrix} 16.390 \\ 8.610 \\ -5 \end{bmatrix} \text{ m/sec.} $$

4.5 Serial Chain of Three Rigid Bodies

Figure 4.13 shows three rigid bodies (body 0, body 1, and body 2). The velocity state of body 1 is known relative to body 0, i.e., $\{{}^0\boldsymbol{\omega}^1;\ {}^0\boldsymbol{v}_O^1\}$, and the velocity state of body 2 is known relative to body 1, i.e., $\{{}^1\boldsymbol{\omega}^2;\ {}^1\boldsymbol{v}_O^2\}$. Note that the coordinate system embedded in body 1 is coincident and aligned with the coordinate system in body 0. In other words, the same reference point O, the origin of the reference frames in body 0 and body 1, is chosen to describe both velocity states. The objective here is to determine the velocity state of body 2 with respect to body 0, i.e., $\{{}^0\boldsymbol{\omega}^2;\ {}^0\boldsymbol{v}_O^2\}$.

Let β be any arbitrary vector that is fixed in body 2. From (4.12) of case 3 of Section 4.2, the time derivative of β with respect to body 0 and body 1 may be written as

$$ \frac{{}^0d}{dt}\beta = {}^0\boldsymbol{\omega}^2 \times {}^0\beta, \qquad (4.60) $$

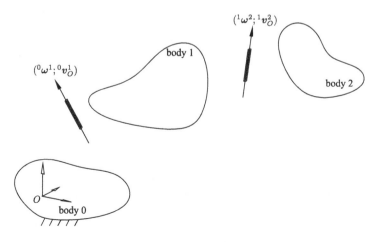

Figure 4.13 Serial chain of three rigid bodies

$$\frac{^1d}{dt}\beta = {}^1\omega^2 \times {}^1\beta. \tag{4.61}$$

From (4.34), these two time derivatives are related by

$$\frac{^0d}{dt}\beta = \frac{^1d}{dt}{}^1\beta + {}^0\omega^1 \times {}^0\beta. \tag{4.62}$$

Substituting (4.60) and (4.61) into (4.62) gives

$${}^0\omega^2 \times {}^0\beta = {}^1\omega^2 \times {}^1\beta + {}^0\omega^1 \times {}^0\beta. \tag{4.63}$$

Since the coordinate systems attached to bodies 0 and 1 are coincident and aligned, ${}^0\beta = {}^1\beta$ and the superscript will be omitted. Rearranging (4.63) yields

$${}^0\omega^2 \times \beta = \left({}^0\omega^1 + {}^1\omega^2\right) \times \beta. \tag{4.64}$$

Since this equation must be valid for any arbitrary vector β, it must be the case that

$${}^0\omega^2 = {}^0\omega^1 + {}^1\omega^2, \tag{4.65}$$

and thus the first velocity state parameter is known in terms of the given quantities.

To determine the second velocity state component, i.e., ${}^1v_O^2$, consider the points A, fixed in body 0, and B, fixed in body 1 (see Figure 4.14). It is apparent from the figure that

$$\boldsymbol{p}_{A0 \to O2} = \boldsymbol{q}_{B1 \to O2} + \boldsymbol{r}_{A0 \to B1}, \tag{4.66}$$

where $\boldsymbol{p}_{A0 \to O2}$ is the position vector from point A in body 0 to point O in body 2 (i.e., the point in body 2 that is at this instant coincident with the origin of the reference frame attached to body 0), $\boldsymbol{q}_{B1 \to O2}$ is the position vector from point B in body 1 to

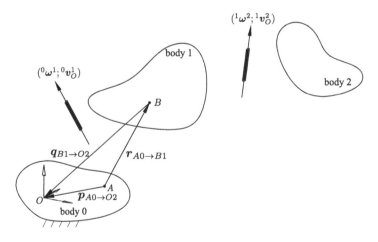

Figure 4.14 Serial chain of three rigid bodies

point O in body 2, and $r_{A0 \to B1}$ is the position vector from point A in body 0 to point B in body 1. The time derivative of (4.66) with respect to body 0 may be written as

$$\frac{^0d}{dt} p_{A0 \to O2} = \frac{^0d}{dt} q_{B1 \to O2} + \frac{^0d}{dt} r_{A0 \to B1}. \tag{4.67}$$

From case 2 of Section 4.2, the left side of this equation may be written as

$$\frac{^0d}{dt} p_{A0 \to O2} = {^0v_O^2}, \tag{4.68}$$

which is the velocity state parameter that is to be obtained. Similarly, the last term in (4.67) can be written as

$$\frac{^0d}{dt} r_{A0 \to B1} = {^0v_B^1}. \tag{4.69}$$

From case 4 of Section 4.2, the first term on the right side of (4.67) may be written as

$$\frac{^0d}{dt} q_{B1 \to O2} = \frac{^1d}{dt} q_{B1 \to O2} + {^0\omega^1} \times q_{B1 \to O2}. \tag{4.70}$$

The term $\frac{^1d}{dt} q_{B1 \to O2}$ is interpreted as the velocity of point O in body 2 as measured with respect to body 1, and thus

$$\frac{^1d}{dt} q_{B1 \to O2} = {^1v_O^2}. \tag{4.71}$$

Substituting (4.71) into (4.70) and then (4.68), (4.69), and (4.70) into equation (4.67) and rearranging gives

$$^0v_O^2 = {^0v_B^1} + {^0\omega^1} \times q_{B1 \to O2} + {^1v_O^2}. \tag{4.72}$$

From (4.41), the velocity of point O in body 1, $^0v_O^1$, may be written as

$$^0v_O^1 = {^0v_B^1} + {^0\omega^1} \times {^0r_{B1 \to O1}}, \tag{4.73}$$

where the term $r_{B1 \to O1}$ represents the vector from point B in body 1 to point O in body 1, which at this instant is equal to $q_{B1 \to O2}$, the vector from point B in body 1 to point O in body 2. Substituting $r_{B1 \to O1} = q_{B1 \to O2}$ into (4.73) and rearranging gives

$$^0v_B^1 = {^0v_O^1} - {^0\omega^1} \times q_{B1 \to O2}. \tag{4.74}$$

Lastly, substituting (4.74) into (4.72) gives the final result

$$^0v_O^2 = {^0v_O^1} + {^1v_O^2}. \tag{4.75}$$

This is a significant result. It has been shown in equations (4.65) and (4.75) that the velocity state of body 2 with respect to body 0 is equal to the sum of the velocity state of body 1 with respect to body 0 and the velocity state of body 2 with respect to body 1. This can be expressed as

$$\begin{bmatrix} ^0\omega^2 \\ ^0v_O^2 \end{bmatrix} = \begin{bmatrix} ^0\omega^1 \\ ^0v_O^1 \end{bmatrix} + \begin{bmatrix} ^1\omega^2 \\ ^1v_O^2 \end{bmatrix}. \tag{4.76}$$

Using screw notation, this equation may be written as

$$^0\hat{T}^2 = {}^0\hat{T}^1 + {}^1\hat{T}^2,\tag{4.77}$$

where

$$^i\hat{T}^j = \begin{bmatrix} ^i\boldsymbol{\omega}^j \\ ^i\boldsymbol{v}_O^j \end{bmatrix} = {}_i\omega_j \begin{bmatrix} ^i\boldsymbol{S}^j \\ ^i\boldsymbol{S}_O^j \end{bmatrix}\tag{4.78}$$

and where $^i\boldsymbol{S}^j$ is a unit vector.

4.6   Serial Chain of Multiple Rigid Bodies

Figure 4.15 shows a serial chain of rigid bodies numbered 0 through n. It is assumed that the velocity state of body $i + 1$ is known relative to body i, $i = 0 \ldots n - 1$. The objective is to determine the velocity state of body n relative to body 0.

It can readily be proven by induction from the results of the previous section that the velocity state of body n relative to body 0 can be calculated from the given velocity states as

$$\begin{bmatrix} ^0\boldsymbol{\omega}^n \\ ^0\boldsymbol{v}_O^n \end{bmatrix} = \begin{bmatrix} ^0\boldsymbol{\omega}^1 \\ ^0\boldsymbol{v}_O^1 \end{bmatrix} + \begin{bmatrix} ^1\boldsymbol{\omega}^2 \\ ^1\boldsymbol{v}_O^2 \end{bmatrix} + \cdots + \begin{bmatrix} ^{n-1}\boldsymbol{\omega}^n \\ ^{n-1}\boldsymbol{v}_O^n \end{bmatrix}.\tag{4.79}$$

Again, for this case, the relative velocity states, i.e., $\{^i\boldsymbol{\omega}^{i+1}; {}^i\boldsymbol{v}_O^{i+1}\}$, are all calculated in terms of the common reference point O. Thus, the coordinate system attached to

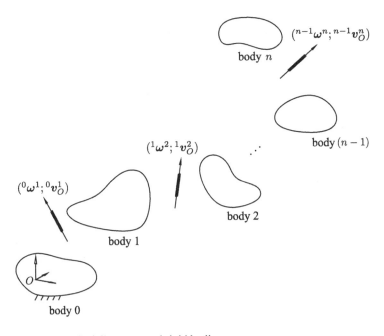

Figure 4.15 n Serially connected rigid bodies

body i in which the velocity state $\left\{ {}^{i}\boldsymbol{\omega}^{i+1};\ {}^{i}\boldsymbol{v}_{O}^{i+1} \right\}$ is measured has its origin at O and its coordinate axes parallel to the coordinate systems attached to all the other bodies.

4.7 Forward and Reverse Velocity Analyses for Serial Manipulators

A non-redundant serial manipulator is defined as a device whose last link, or end effector, has six degrees of freedom. Thus, the sum of the freedoms of the individual joints of the manipulator is six. A revolute, prismatic, or helical joint introduces one degree of freedom, while a cylindric joint introduces two degrees of freedom.

For the forward and reverse velocity analyses, it is assumed that the position analysis for the serial manipulator has been completed. In other words, the individual joint parameters are known for the manipulator. Thus, it is assumed that the Plücker line coordinates can be determined along each and every joint axis of the manipulator. Furthermore, it is assumed that the pitch along each joint axis is known (for revolute joints this pitch will equal zero). Each link of the manipulator is, thus, a rigid body, and these links are connected in series by the joints whose pitch and lines of action are known.

Figure 4.16 shows a kinematic diagram of a non-redundant six-axis robot manipulator with joint axis vectors ${}^{0}\boldsymbol{S}^{1}$ through ${}^{5}\boldsymbol{S}^{6}$ labeled. In this example, the first three joints are revolute joints, the fourth joint is a prismatic joint, the fifth joint is a helical joint with pitch h, and the sixth joint is another revolute joint. Since it is assumed that the position analysis is completed, the directions of the unit vectors ${}^{i}\boldsymbol{S}^{i+1}$ along each axis as well the coordinates of one point on each joint axis are assumed to be known. From this information, the Plücker coordinates of the lines along the revolute and helical joint axes are known as

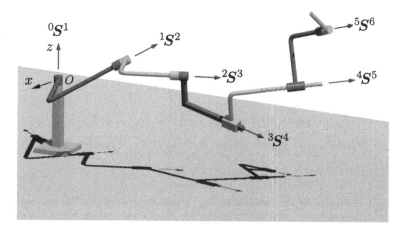

Figure 4.16 Kinematic diagram of serial robot manipulator

$$\{{}^i S^{i+1};\, {}^i S_{OL}^{i+1}\} = \{{}^i S^{i+1};\, r_i \times {}^i S^{i+1}\}, \tag{4.80}$$

where r_i is a vector from the reference point O to a point on the corresponding joint axis. The coordinates of the line at infinity associated with the prismatic joint may be written as $\{0\,;\,{}^3 S^4\}$, where the components of the unit direction vector ${}^3 S^4$ are dimensionless.

In the forward velocity analysis, it is assumed that the angular velocity for each revolute and helical joint and the linear velocity for each prismatic joint are known. Thus, for this case it is assumed that the scalar values ${}_0\omega_1$, ${}_1\omega_2$, ${}_2\omega_3$, ${}_3v_4$, ${}_4\omega_5$, and ${}_5\omega_6$ are known. The objective is to determine the velocity state of the last link of the manipulator relative to a reference frame attached to the base. From (4.79), it is apparent that the velocity state of the last link will equal the sum of the individual relative velocity states. For the particular case of the manipulator shown in Figure 4.16, this may be written as

$$\begin{bmatrix} {}^0\omega^6 \\ {}_0v_O^6 \end{bmatrix} = {}_0\omega_1 \,{}^0\left(\begin{bmatrix} {}^0 S^1 \\ {}^0 S_{OL}^1 \end{bmatrix} \right) + {}_1\omega_2 \,{}^0\left(\begin{bmatrix} {}^1 S^2 \\ {}^1 S_{OL}^2 \end{bmatrix} \right) + {}_2\omega_3 \,{}^0\left(\begin{bmatrix} {}^2 S^3 \\ {}^2 S_{OL}^3 \end{bmatrix} \right) \tag{4.81}$$

$$+ {}_3v_4 \,{}^0\left(\begin{bmatrix} 0 \\ {}^3 S^4 \end{bmatrix} \right) + {}_4\omega_5 \,{}^0\left(\begin{bmatrix} {}^4 S^5 \\ {}^4 S_{OL}^5 + {}_4 h_5 \,{}^4 S^5 \end{bmatrix} \right) + {}_5\omega_6 \,{}^0\left(\begin{bmatrix} {}^5 S^6 \\ {}^5 S_{OL}^6 \end{bmatrix} \right).$$

The velocity state for each revolute joint is simply the joint angular velocity times the Plücker line coordinates of the revolute axis. For the helical axis, a pitch term is added as per (4.59). For the prismatic joint, all points in body 4, including the reference point, have zero angular velocity and move along the direction of ${}^3 S^4$ with a velocity of ${}_3v_4$. Note that the coordinates of all the screws are written in terms of the fixed base coordinate system.

All terms on the right side of (4.81) are known quantities and, thus, the forward velocity analysis, that is the determination of the velocity state of body 6 with respect to body 0, is a straightforward task. For the reverse velocity analysis, the desired velocity state of body 6 relative to body 0 is given, i.e., $\{{}^0\omega^6; {}^0v_O^6\}$, and it is necessary to compute the scalar joint velocity parameters, which for this case are ${}_0\omega_1$, ${}_0\omega_1$, ${}_1\omega_2$, ${}_2\omega_3$, ${}_3v_4$, and ${}_5\omega_6$. Equation (4.81) can be expanded as six scalar equations in the six unknown velocity parameters. The reverse velocity analysis is thus completed by solving these six equations in six unknowns.[1]

In general, for the case of all revolute joints, which is common for many industrial manipulators, equation (4.79) may be written as

$$\begin{bmatrix} {}^0\omega^6 \\ {}_0v_O^6 \end{bmatrix} = {}_0\omega_6 \begin{bmatrix} {}^0 S^6 \\ {}^0 S_O^6 \end{bmatrix} = {}_0\omega_6 \,{}^0\$^6 \tag{4.82}$$

$$= {}_0\omega_1 \,{}^0\$^1 + {}_1\omega_2 \,{}^1\$^2 + {}_2\omega_3 \,{}^2\$^3 + {}_3\omega_4 \,{}^3\$^4 + {}_4\omega_5 \,{}^4\$^5 + {}_5\omega_6 \,{}^5\$^6,$$

---

[1] Two relative velocity terms would be used for the case of a cylindric joint. The first would represent a rotation about the line along the joint axis and the second would represent a translation in the direction of the joint axis.

where $^i\$^{i+1}$ are the appropriate unitized screw coordinates for each joint axis. Equation (4.82) may then be written in the matrix format

$$\begin{bmatrix} {}^0\omega^6 \\ {}^0v_O^6 \end{bmatrix} = {}_0\omega_6\ {}^0\$^6 = \boldsymbol{J}\ \omega, \tag{4.83}$$

where $\boldsymbol{J}$, named the Jacobian matrix, is the 6×6 matrix formed from the unitized screw coordinates (line coordinates for revolute joints) as

$$\boldsymbol{J} = \begin{bmatrix} {}^0\$^1 & {}^1\$^2 & {}^2\$^3 & {}^3\$^4 & {}^4\$^5 & {}^5\$^6 \end{bmatrix} \tag{4.84}$$

$$= \begin{bmatrix} {}^0S^1 & {}^1S^2 & {}^2S^3 & {}^3S^4 & {}^4S^5 & {}^5S^6 \\ {}^0S_{OL}^1 & {}^1S_{OL}^2 & {}^2S_{OL}^3 & {}^3S_{OL}^4 & {}^4S_{OL}^5 & {}^5S_{OL}^6 \end{bmatrix}$$

and ω is the vector of joint velocities given by

$$\omega = \begin{bmatrix} {}_0\omega_1 \\ {}_1\omega_2 \\ {}_2\omega_3 \\ {}_3\omega_4 \\ {}_4\omega_5 \\ {}_5\omega_6 \end{bmatrix}. \tag{4.85}$$

In summary, since the position analysis is assumed to be completed, all components of the Jacobian matrix $\boldsymbol{J}$ will be known. The forward velocity analysis for the manipulator is then completed by solving (4.83) for the velocity state of body 6 with respect to body 0, since the elements of vector ω are given. For the reverse velocity analysis, the Jacobian matrix $\boldsymbol{J}$ is again known together with the desired velocity state of body 6 relative to body 0. The individual joint velocities can then be obtained from

$$\omega = \boldsymbol{J}^{-1} \begin{bmatrix} {}^0\omega^6 \\ {}^0v_O^6 \end{bmatrix}. \tag{4.86}$$

It is possible that the manipulator could be in a position configuration in which the Jacobian matrix is singular and, thus, its inverse cannot be obtained. Such configurations are referred to as singular configurations, and these will be discussed in Chapter 6.

4.7.1   Example Problem

Shown in Figure 4.17 is a drawing of a Puma robot, and a kinematic model of this robot is shown in Figure 4.18. This manipulator is comprised of six revolute joints such that the first two joint axes intersect, the second and third are parallel, and the last three intersect. The constant mechanism parameters for this manipulator, based on the directions of the joint axis and link vectors shown in Figure 4.18, are shown in Table 4.1 (the link vectors $\boldsymbol{a}^4$ and $\boldsymbol{a}^5$ are not shown in the figure, but their directions can be deduced based on the given mechanism parameters). For this analysis, the joint offset ${}_5S_6$ was chosen to equal 4 in.

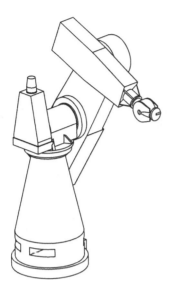

Figure 4.17 Puma robot

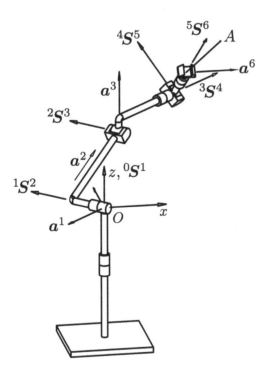

Figure 4.18 Kinematic model of puma robot

Table 4.1. Constant mechanism parameters for Puma manipulator

Link Length, in	Twist Angle, deg	Joint Offset, in
$a_1 = 0$	$\alpha_1 = 90$	
$a_2 = 17$	$\alpha_2 = 0$	$_1S_2 = 5.9$
$a_3 = 0.8$	$\alpha_3 = 270$	$_2S_3 = 0$
$a_4 = 0$	$\alpha_4 = 90$	$_3S_4 = 17$
$a_5 = 0$	$\alpha_5 = 90$	$_4S_5 = 0$

At the instant shown, the joint angle values are given as

$$_0\phi_1 = 225 \deg$$
$$_1\theta_2 = 150 \deg$$
$$_2\theta_3 = -60 \deg$$
$$_3\theta_4 = 45 \deg$$
$$_4\theta_5 = 60 \deg$$
$$_5\theta_6 = -30 \deg .$$

The coordinates of point A, the tool point, are given in terms of the end effector coordinate system as

$$^6P_A^6 = \begin{bmatrix} 5 \\ 3 \\ 7 \end{bmatrix} \text{ in .}$$

It is desired to obtain the six individual joint angular velocities such that the last link of the manipulator will have an angular velocity of $^0\omega^6 = [0.20, -0.12, 0.10]^T$ rad/sec and point A will have a velocity of $^0v_A^6 = [6.5, 3.5, -4.0]^T$ in/sec.

The solution to this problem will proceed in three steps. In the first, the desired velocity state of the last link of the manipulator, that is $\{^0\omega^6; ^0v_O^6\}$, will be determined. Then, the Jacobian matrix defined by (4.84) will be computed based on the given values. Finally, the six joint angular velocities will be computed by inverting the Jacobian matrix and applying equation (4.86). The first step begins by performing a forward position analysis to determine the coordinates of point A measured in terms of the fixed coordinate system. With this information, it will be possible to obtain the linear velocity of the point in body 6, which at this instant is coincident with the origin of the fixed coordinate system, i.e., $^0v_O^6$, and, thus, the velocity state parameters of link 6, $\{^0\omega^6; ^0v_O^6\}$ will be known.

Equations (2.80) and (2.79) are used to obtain the 4×4 transformation matrices $_1^FT$ through $_{i+1}^iT, i = 0 \ldots 5$ as

$$_1^FT = \begin{bmatrix} -0.7071 & 0.7071 & 0 & 0 \\ -0.7071 & -0.7071 & 0 & 0 \\ 0 & 0 & 1 & 0 \\ 0 & 0 & 0 & 1 \end{bmatrix}$$

$$
{}^1_2T = \begin{bmatrix} -0.8660 & -0.5 & 0 & 0 \\ 0 & 0 & -1 & -5.9 \\ 0.5 & -0.8660 & 0 & 0 \\ 0 & 0 & 0 & 1 \end{bmatrix}
$$

$$
{}^2_3T = \begin{bmatrix} 0.5 & 0.8660 & 0 & 17 \\ -0.8660 & 0.5 & 0 & 0 \\ 0 & 0 & 1 & 0 \\ 0 & 0 & 0 & 1 \end{bmatrix}
$$

$$
{}^3_4T = \begin{bmatrix} 0.7071 & -0.7071 & 0 & 0.8 \\ 0 & 0 & 1 & 17 \\ -0.7071 & 0.7071 & 0 & 0 \\ 0 & 0 & 0 & 1 \end{bmatrix}
$$

$$
{}^4_5T = \begin{bmatrix} 0.5 & -0.8660 & 0 & 0 \\ 0 & 0 & -1 & 0 \\ 0.8660 & 0.5 & 0 & 0 \\ 0 & 0 & 0 & 1 \end{bmatrix}
$$

$$
{}^5_6T = \begin{bmatrix} 0.8660 & 0.5 & 0 & 0 \\ 0 & 0 & -1 & -4 \\ -0.5 & 0.8660 & 0 & 0 \\ 0 & 0 & 0 & 1 \end{bmatrix}.
$$

The terms of the transformation matrices are dimensionless with the exception of the first three elements of the fourth column, which have units of inches. The transformation matrices ${}^F_iT$, $i = 2 \ldots 6$ are obtained as

$$
{}^F_iT = {}^F_1T\ {}^1_2T \cdots {}^{i-1}_iT,\ i = 2 \ldots 6
$$

and the numerical results are

$$
{}^F_2T = \begin{bmatrix} 0.6124 & 0.3536 & -0.7071 & -4.172 \\ 0.6124 & 0.3536 & 0.7071 & 4.1712 \\ 0.5 & -0.8660 & 0 & 0 \\ 0 & 0 & 0 & 1 \end{bmatrix}
$$

$$
{}^F_3T = \begin{bmatrix} 0 & 0.7071 & -0.7071 & 6.238 \\ 0 & 0.7071 & 0.7071 & 14.582 \\ 1 & 0 & 0 & 8.5 \\ 0 & 0 & 0 & 1 \end{bmatrix}
$$

$$
{}_4^F T = \begin{bmatrix} 0.5 & 0.5 & 0.7071 & 18.259 \\ -0.5 & -0.5 & 0.7071 & 26.603 \\ 0.7071 & -0.7071 & 0 & 9.3 \\ 0 & 0 & 0 & 1 \end{bmatrix}
$$

$$
{}_5^F T = \begin{bmatrix} 0.8624 & -0.0795 & -0.5 & 18.259 \\ 0.3624 & 0.7866 & 0.5 & 26.603 \\ 0.3536 & -0.6124 & 0.7071 & 9.3 \\ 0 & 0 & 0 & 1 \end{bmatrix}
$$

$$
{}_6^F T = \begin{bmatrix} 0.9968 & -0.0018 & 0.0795 & 18.577 \\ 0.0638 & 0.6142 & -0.7866 & 23.457 \\ -0.0474 & 0.7891 & 0.6124 & 11.750 \\ 0 & 0 & 0 & 1 \end{bmatrix}.
$$

The coordinates of the tool point, point A, may now be obtained in terms of the fixed reference frame as

$$
{}^0 P_A^6 = {}_6^F T \; {}^6 P_A^6 = \begin{bmatrix} 24.112 \\ 20.113 \\ 18.167 \end{bmatrix} \text{ in,}
$$

which is written here in Cartesian coordinates rather than homogeneous coordinates. Equation (4.41) can be used to determine the velocity of point O in body 6 as

$$
{}^0 v_O^6 = {}^0 v_A^6 - {}^0 \omega^6 \times {}^0 r_{O \to A} = \begin{bmatrix} 10.6913 \\ 4.7221 \\ -10.9160 \end{bmatrix} \text{ in/sec.}
$$

The next step is to determine the coordinates of the lines along the six revolute joints. The directions of the six joint axis vectors expressed in terms of the fixed reference coordinate system shown in the figure are obtained as the first three elements of the third column of the matrices ${}_i^F T$, $i = 1 \dots 6$ and are written as

$$
{}^0 \left({}^0 S^1 \right) = \begin{bmatrix} 0 \\ 0 \\ 1 \end{bmatrix}, \quad {}^0 \left({}^1 S^2 \right) = \begin{bmatrix} -0.7071 \\ 0.7071 \\ 0 \end{bmatrix}, \quad {}^0 \left({}^2 S^3 \right) = \begin{bmatrix} -0.7071 \\ 0.7071 \\ 0 \end{bmatrix},
$$

$$
{}^0 \left({}^3 S^4 \right) = \begin{bmatrix} 0.7071 \\ 0.7071 \\ 0 \end{bmatrix}, \quad {}^0 \left({}^4 S^5 \right) = \begin{bmatrix} -0.5 \\ 0.5 \\ 0.7071 \end{bmatrix}, \quad {}^0 \left({}^5 S^6 \right) = \begin{bmatrix} 0.0795 \\ -0.7866 \\ 0.6124 \end{bmatrix}.
$$

The preceding superscript 0 is used to indicate that each of the vectors has been evaluated in terms of the fixed coordinate system attached to body 0. A point on the ith joint axes can be obtained as the first three elements of the fourth column of the matrix ${}_i^F T$, $i = 1 \dots 6$, since this represents the coordinates of the origin of the standard ith coordinate system as measured in the fixed reference coordinate system. Thus, the moment of the line along the ith joint axis can be calculated as

$$^{i-1}S^i_{OL} = {}^F P^i_{iorig} \times {}^{i-1}S^i, \; i = 1 \cdots 6.$$

The numerical results are

$$^0 \left({}^0S^1_{OL} \right) = \begin{bmatrix} 0 \\ 0 \\ 0 \end{bmatrix}, \qquad {}^0 \left({}^1S^2_{OL} \right) = \begin{bmatrix} 0 \\ 0 \\ 0 \end{bmatrix}, \qquad {}^0 \left({}^2S^3_{OL} \right) = \begin{bmatrix} -6.0104 \\ -6.0104 \\ 14.7224 \end{bmatrix},$$

$$^0 \left({}^3S^4_{OL} \right) = \begin{bmatrix} -6.5761 \\ 6.5761 \\ -5.9000 \end{bmatrix}, \quad {}^0 \left({}^4S^5_{OL} \right) = \begin{bmatrix} 14.1612 \\ -17.5612 \\ 22.4311 \end{bmatrix}, \quad {}^0 \left({}^5S^6_{OL} \right) = \begin{bmatrix} 23.6061 \\ -10.4425 \\ -16.4759 \end{bmatrix},$$

where all terms have units of inches. The Jacobian matrix can now be written as

$$J = \begin{bmatrix} 0 & -0.7071 & -0.7071 & 0.7071 & -0.5 & 0.0795 \\ 0 & 0.7071 & 0.7071 & 0.7071 & 0.5 & -0.7866 \\ 1 & 0 & 0 & 0 & 0.7071 & 0.6124 \\ 0 & 0 & -6.0104 & -6.5761 & 14.1612 & 23.6061 \\ 0 & 0 & -6.0104 & 6.5761 & -17.5612 & -10.4425 \\ 0 & 0 & 14.7224 & -5.9000 & 22.4311 & -16.4759 \end{bmatrix},$$

where the elements of the first three rows of J are dimensionless and the elements of the last three rows have units of inches. Finally, from (4.86), the six joint angular velocities are calculated as

$$\omega = J^{-1} \begin{bmatrix} {}^0\omega^6 \\ {}^0v^6_O \end{bmatrix} = \begin{bmatrix} -0.0496 \\ 0.9929 \\ -1.6525 \\ -0.0592 \\ 0.4122 \\ -0.2316 \end{bmatrix} \text{ rad/sec.}$$

4.7.2   Example Problem

Figure 4.19 shows a closed-loop mechanism where body 0 is fixed. This is a one degree of freedom mechanism due to the special geometry where the axes of the three revolute joints are parallel and the direction of the prismatic joint is in a plane that is perpendicular to the direction of the revolute joint axes. For this problem, the magnitude of the angular velocity of body 1 relative to body 0 is given as $_0\omega_1 = 0.25$ rad/sec in the direction shown in the figure. The objective is to determine the angular velocity of body 3 relative to body 0, i.e., $_0\omega_3$.

The directions of the unit vectors $^i S^{i+1}$ may be written as

$$^0S^1 = {}^1S^2 = {}^3S^0 = \begin{bmatrix} 0 \\ 0 \\ 1 \end{bmatrix}, \quad {}^2S^3 = \begin{bmatrix} \cos 45^\circ \\ \sin 45^\circ \\ 0 \end{bmatrix}.$$

The relative velocity states of the bodies, as measured in the coordinate system shown in the figure, are added together to obtain the velocity state of body 0 relative to itself to yield

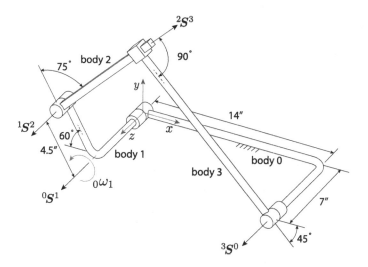

Figure 4.19 Closed-loop mechanism

$$_0\omega_1 \begin{bmatrix} {}^0S^1 \\ 0 \end{bmatrix} + {}_1\omega_2 \begin{bmatrix} {}^1S^2 \\ r_1 \times {}^1S^2 \end{bmatrix} + {}_2v_3 \begin{bmatrix} 0 \\ {}_2S^3 \end{bmatrix} + {}_3\omega_0 \begin{bmatrix} {}^3S^0 \\ r_3 \times {}^3S^0 \end{bmatrix} = \begin{bmatrix} 0 \\ 0 \end{bmatrix}.$$

The vector r_1 is a vector to any point on the axis of rotation of body 2 relative to body 1 and the vector r_3 is a vector to any point on the axis of rotation of body 0 relative to body 3. These vectors may be written as

$$r_1 = \begin{bmatrix} -4.5\cos 60° \\ 4.5\sin 60° \\ 7 \end{bmatrix}, \quad r_3 = \begin{bmatrix} 14 \\ 0 \\ 7 \end{bmatrix} \text{ in.}$$

The sum of the velocity states may now be written as

$$_0\omega_1 \begin{bmatrix} 0 \\ 0 \\ 1 \\ 0 \\ 0 \\ 0 \end{bmatrix} + {}_1\omega_2 \begin{bmatrix} 0 \\ 0 \\ 1 \\ 3.8971 \\ 2.25 \\ 0 \end{bmatrix} + {}_2v_3 \begin{bmatrix} 0 \\ 0 \\ 0 \\ 0.7071 \\ 0.7071 \\ 0 \end{bmatrix} + {}_3\omega_0 \begin{bmatrix} 0 \\ 0 \\ 1 \\ 0 \\ -14 \\ 0 \end{bmatrix} = \begin{bmatrix} 0 \\ 0 \\ 0 \\ 0 \\ 0 \\ 0 \end{bmatrix}.$$

Note that the top three rows will have units of rad/sec while the bottom three rows will have units of in/sec. It can be seen that in this case the first two rows and the last row all contain zeros. The three remaining equations can be written as

$$_0\omega_1 \begin{bmatrix} 1 \\ 0 \\ 0 \end{bmatrix} + {}_1\omega_2 \begin{bmatrix} 1 \\ 3.8971 \\ 2.25 \end{bmatrix} + {}_2v_3 \begin{bmatrix} 0 \\ 0.7071 \\ 0.7071 \end{bmatrix} + {}_3\omega_0 \begin{bmatrix} 1 \\ 0 \\ -14 \end{bmatrix} = \begin{bmatrix} 0 \\ 0 \\ 0 \end{bmatrix}.$$

Substituting $_0\omega_1 = 0.25$ rad/sec and solving for the remaining terms yields

$$_1\omega_2 = -0.2833 \text{ rad/sec}$$
$$_2v_3 = 1.5616 \text{ in/sec}$$
$$_3\omega_0 = 0.0333 \text{ rad/sec.}$$

The magnitude of the angular velocity of body 3 relative to body 0 will be the negative of $_3\omega_0$ and thus

$$_0\omega_3 = -0.0333 \text{ rad/sec.}$$

4.8   Problems

1. The angular velocity of body 1 measured with respect to body 0 is given as $^0\omega^1 = [2, 0.5, -1.5]^T$ rad/sec. At this instant, the coordinates of a point P, which is embedded in body 1, is measured with respect to body 0 as $[-1, 2, 0.5]^T$ m, and the velocity of this point measured with respect to body 0 is $[10, 10, 0]^T$ m/sec.

   (a) Determine the velocity of the point in body 1 that is at this instant coincident with the origin of the body 0 coordinate system, and write the coordinates of the instantaneous twist $_0\omega_1 {}^0\$^1$.
   (b) Determine the pitch of the twist.
   (c) Determine the Plücker coordinates of the line of action of the twist.
   (d) Determine the coordinates of the point on the line of action that is closest to the origin of the body 0 coordinate system.

2. Body 1 is rotating about the line $\{1, 1, 1; 2, -1, -1\}$, where the first three terms are dimensionless and the last three have units of meters, with an angular velocity of 2.5 rad/sec. It is also translating in the direction $[2, 1, 1]^T$ at a speed of 4 m/sec with respect to ground (body 0). At this instant, body 2 is rotating about the screw $\{-1, 2, 4; 4, -4, 10\}$, where the first three terms are dimensionless and the last three have units of meters, with an angular velocity of 1.5 rad/sec with respect to body 1. At this instant, the coordinate system attached to body 1 is aligned and coincident with the coordinate system attached to body 0. Determine the instantaneous twist that body 2 is moving on with respect to ground.

3. The kinematic model of the Cincinnati Milacron T3-776 robot manipulator is shown in Figure 4.20. The constant mechanism parameters for this kinematic model are given in Table 4.2.

   The sixth coordinate system is now defined by selecting the length of the offset $_5S_6$ to be six inches. The coordinates of the tool point attached to the last link of the robot is then determined to be $[5, 3, 7]^T$ inches, as measured in terms of the sixth coordinate system.

Table 4.2. Constant mechanism parameters for T3-776 manipulator

Link Length, in	Twist Angle, deg	Joint Offset, in
$a_1 = 0$	$\alpha_1 = 90$	
$a_2 = 44$	$\alpha_2 = 0$	$_1S_2 = 0$
$a_3 = 0$	$\alpha_3 = 90$	$_2S_3 = 0$
$a_4 = 0$	$\alpha_4 = 61$	$_3S_4 = 55$
$a_5 = 0$	$\alpha_5 = 61$	$_4S_5 = 0$

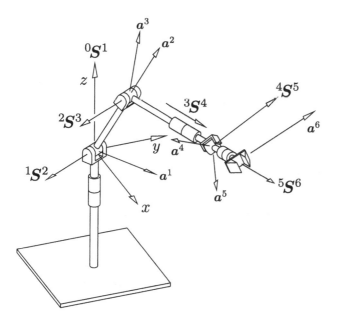

Figure 4.20 Kinematic model of Cincinnati Milacron T3-776 manipulator

At a particular instant of time, the joint angle parameters of the robot are measured in deg as

$$_0\phi_1 = 37°$$
$$_1\theta_2 = 85°$$
$$_2\theta_3 = -23°$$
$$_3\theta_4 = 71°$$
$$_4\theta_5 = 127°$$
$$_5\theta_6 = 101°.$$

At this instant, it is desired that the tool point has a velocity of $[2, 4, -7]^T$ in/sec and that the last link of the robot has an angular velocity of $[8, 0, 0]^T$ deg/sec (make sure to convert to rad/sec) relative to the fixed coordinate system.

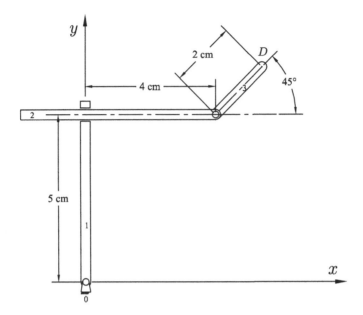

Figure 4.21 RPR manipulator

Determine the angular velocities for each of the joints in order that the end effector moves as desired at this instant.

4. A planar RPR manipulator is shown in Figure 4.21.
   (a) The following information is given for the instant shown in the figure:
       (i) the angular velocity of link 1 measured with respect to ground (body 0) is given as $_0\omega_1 = 1.5$ rad/sec (direction is counterclockwise)
       (ii) the linear velocity of link 2 measured with respect to link 1 is given as $_1v_2 = 3$ cm/sec (direction to the right)
       (iii) the angular velocity of link 3 measured with respect to link 2 is given as $_2\omega_3 = -2$ rad/sec (direction is clockwise).
       Determine the velocity state of body 3 at this instant measured with respect to body 0. Determine the velocity of point D measured with respect to ground.
   (b) At the instant shown in Figure 4.21, it is desired that the end effector translate at a speed of 4 cm/sec in a direction parallel to the vector $3i + 4j$. Determine values for the joint parameter velocities $_0\omega_1$, $_1v_2$, and $_2\omega_3$ that will cause the end effector to translate as desired, if possible.

5. The mechanism shown in Figure 4.22 is a two degree of freedom device. Actuators are attached that can move the two sliding blocks, i.e., bodies 1 and 4, relative to ground (body 0). At the instant shown, the coordinates of points A, B, C, and D are given as

$$p_A = \begin{bmatrix} 0 \\ 50 \\ 0 \end{bmatrix} \text{in}, \qquad p_B = \begin{bmatrix} 45 \\ 30 \\ 0 \end{bmatrix} \text{in}, \qquad p_C = \begin{bmatrix} 30 \\ 0 \\ 0 \end{bmatrix} \text{in}, \qquad p_D = \begin{bmatrix} 30 \\ 60 \\ 0 \end{bmatrix} \text{in}$$

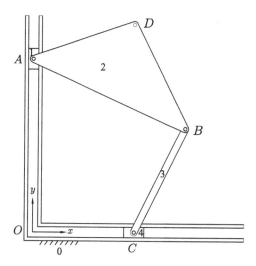

Figure 4.22 PRRRP planar mechanism

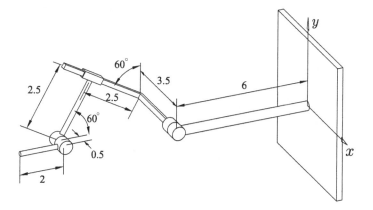

Figure 4.23 RPR manipulator

   (a) The velocity of body 1 measured with respect to ground is given as
       $[0, -6, 0]^T$ in/sec, and the velocity of body 4 measured with respect to ground
       is given as $[4, 0, 0]^T$ in/sec. Determine the velocity of point D at this instant.
   (b) Determine the linear velocity of the two sliding blocks that would cause point
       D to move at this instant in the y direction with a velocity of 5 in/sec while
       body 2 has an angular velocity of 0.25 rad/sec about the z axis relative to
       body 0.
6. A RPR manipulator is shown in Figure 4.23. Linear dimensions are shown in
   meters. At the instant shown it is desired that the last link be moving in pure
   translation in the Y direction at a speed of 2.5 m/sec. Determine the joint
   velocities that will accomplish this.

5   Reciprocal Screws

Tyger Tyger! burning bright
In the forests of the night,
What immortal hand or eye
Dare frame thy fearful symmetry?

William Blake (1757–1857)
(old "eye" had a long 'E' sound)

5.1   Introduction

The concept of two screws that are reciprocal to one another is introduced. Think of one screw as the instantaneous allowable twist of a rigid body relative to ground and the other as the screw associated with an applied wrench to that body. The two screws are said to be reciprocal if no work is done by the wrench with regards to the allowable twist. In other words, the rigid body will not move about the twist no matter how large the magnitude of the wrench is. It will be shown how this concept of reciprocal screws can be applied to solve velocity problems for serial and parallel robot manipulators.

5.2   Definition of Reciprocal Screws

Figure 5.1 shows a rigid body that is constrained to move on a screw ${}^0\$^1 = \{{}^0S^1; {}^0S_O^1\}$. A wrench of intensity f_2 is applied to the body on a screw $\$_2 = \{S_2; S_{O2}\}$, and it produces a twist of amplitude ${}_0\omega_1$ about the screw ${}^0\$^1$. The pitch of the twist is h_1, while the pitch of the wrench is h_2. It is assumed that the vectors ${}^0S^1$ and S_2 are unit vectors. Also shown in the figure is the unique perpendicular line to the lines of action of the screws ${}^0\$^1$ and $\$_2$. The direction of the vector a_{12} is defined as parallel to $S_2 \times {}^0S^1$, and the angle α_{12} is defined in a right hand sense from S_2 to ${}^0S^1$ about a_{12}. The perpendicular distance between the two lines of action is labeled a_{12}. It should be noted that the coordinates of the points of intersection of the perpendicular line with each of the lines of action can be computed as described in Section 1.11 and, thus, the Plücker coordinates of the mutual perpendicular line are readily determined.

The virtual power that is produced by the wrench about the twist can be computed as the sum of all forces projected upon translational velocities and all torques projected

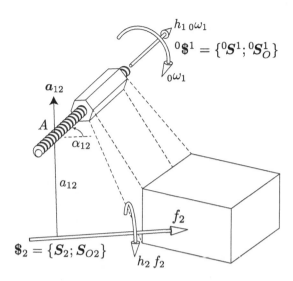

Figure 5.1 A wrench and a twist

upon rotational velocities. For this case, the virtual power that is produced by the wrench can be expressed as the sum of three terms. It can be seen from Figure 5.1 that the moment of the force $f_2 S_2$ about point A is $-a_{12}a_{12} \times f_2 S_2$. The first term is the projection of this moment on the twist axis $^0S^1$ times the angular velocity $_0\omega_1$ about the twist axis. The second is the projection of the force $f_2 S_2$ on the direction of the translational velocity along the twist axis times the magnitude of the translational velocity along the twist axis, which is caused by the pitch of the twist screw. The third is the projection of the moment vector along the axis of the wrench upon the direction of the twist axis times the magnitude of the rotational velocity about the twist axis. The virtual power equation can, thus, be written as

$$U = (-a_{12}a_{12} \times f_2 S_2) \cdot {_0\omega_1}\,{}^0S^1 + (f_2 S_2 \cdot {}^0S^1)(h_1\,{_0\omega_1}) \tag{5.1}$$
$$+ (f_2 h_2 S_2 \cdot {_0\omega_1}\,{}^0S^1).$$

Substituting $^0S^1 \cdot S_2 = \cos \alpha_{12}$ and rearranging yields

$$U = -a_{12}\, f_2\,{_0\omega_1}(a_{12} \times S_2 \cdot {}^0S^1) + f_2\,{_0\omega_1}(h_1 + h_2)\cos \alpha_{12}. \tag{5.2}$$

Rearranging the order of the scalar triple product in the first term of (5.2) and substituting $S_2 \times {}^0S^1 = a_{12} \sin \alpha_{12}$ yields

$$U = f_2\,{_0\omega_1}[(h_1 + h_2)\cos \alpha_{12} - a_{12} \sin \alpha_{12}]. \tag{5.3}$$

If the pitches h_1 and h_2, the twist angle α_{12}, and the mutual perpendicular distance a_{12} are selected so that

$$(h_1 + h_2)\cos \alpha_{12} - a_{12} \sin \alpha_{12} = 0, \tag{5.4}$$

then the wrench cannot produce motion on the screw $_0\$^1$, no matter how large the intensity of f_2, *and the twist and wrench are said to be reciprocal.*

5.3   Reciprocal Product

The reciprocal product of the twist $_0\omega_1\ {}^0\$^1$ and the wrench $f_2\$_2$ is now defined as

$$_0\omega_1\ {}^0\$^1 \circ f_2\$_2 = {}_0\omega_1\ f_2({}^0S^1 \cdot S_{O2} + S_2 \cdot {}^0S_O^1). \tag{5.5}$$

Substituting $S_{O2} = r_2 \times S_2 + h_2 S_2$ and ${}^0S_O^1 = r_1 \times {}^0S^1 + h_1\ {}^0S^1$, where r_1 and r_2 represent any arbitrary point on the line of action of the twist and the wrench, respectively, gives

$$_0\omega_1\ {}^0\$^1 \circ f_2\$_2 = {}_0\omega_1\ f_2[{}^0S^1 \cdot (r_2 \times S_2 + h_2 S_2) + S_2 \cdot (r_1 \times {}^0S^1 + h_1\ {}^0S^1)]. \tag{5.6}$$

Substituting ${}^0S^1 \cdot S_2 = \cos\alpha_{12}$ and rearranging this equation gives

$$_0\omega_1\ {}^0\$^1 \circ f_2\$_2 = {}_0\omega_1\ f_2[h_2\cos\alpha_{12} + {}^0S^1 \cdot r_2 \times S_2 + h_1\cos\alpha_{12} + S_2 \cdot r_1 \times {}^0S^1]. \tag{5.7}$$

Rearranging the order of the two scalar triple products and regrouping yields

$$_0\omega_1\ {}^0\$^1 \circ f_2\$_2 = {}_0\omega_1\ f_2[(h_1 + h_2)\cos\alpha_{12} + {}^0S^1 \cdot (S_2 \times (r_1 - r_2))]. \tag{5.8}$$

The vectors r_1 and r_2 can be replaced by the vectors e_1 and e_2, i.e., the intersection points of the unique perpendicular line with the lines of action of the twist and the wrench, respectively, to give

$$_0\omega_1\ {}^0\$^1 \circ f_2\$_2 = {}_0\omega_1\ f_2[(h_1 + h_2)\cos\alpha_{12} + {}^0S^1 \cdot (S_2 \times (e_1 - e_2))]. \tag{5.9}$$

Substituting $(e_1 - e_2) = a_{12}a_{12}$ and again rearranging the scalar triple product gives

$$_0\omega_1\ {}^0\$^1 \circ f_2\$_2 = {}_0\omega_1\ f_2[(h_1 + h_2)\cos\alpha_{12} - a_{12}a_{12} \cdot (S_2 \times {}^0S^1)]. \tag{5.10}$$

Substituting $S_2 \times {}^0S^1 = a_{12}\sin\alpha_{12}$ and recognizing that $a_{12} \cdot a_{12} = 1$ gives

$$_0\omega_1\ {}^0\$^1 \circ f_2\$_2 = {}_0\omega_1\ f_2[(h_1 + h_2)\cos\alpha_{12} - a_{12}\sin\alpha_{12}]. \tag{5.11}$$

Comparison of (5.3) and (5.11) results in the conclusion that the reciprocal product of the wrench $f_2\$_2$ and the twist $_0\omega_1\ {}^0\$^1$ is equal to the virtual power of the wrench about the twist.

5.3.1   Axes of the Two Screws Intersect

It is informative to consider some special cases of (5.4) where the reciprocal product equals zero. Firstly, suppose the axes of the two screws intersect and, thus, $a_{12} = 0$ (see Figure 5.2). Equation (5.4) reduces to

$$(h_1 + h_2)\cos\alpha_{12} = 0. \tag{5.12}$$

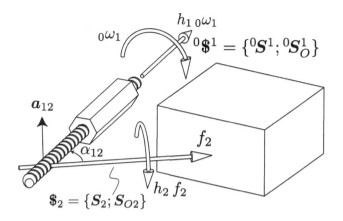

Figure 5.2 A wrench intersecting a twist

For $\cos\alpha_{12} \neq 0$, the two screws are reciprocal when their pitches are equal and opposite, that is, $h_1 = -h_2$. Further, if $h_1 = 0$, then the instant motion about $_0\omega_1 \, ^0\$^1$ is a pure rotation, which can be modeled by a revolute joint. From (5.12), $h_2 = 0$ and the reciprocal wrench is a pure force.

5.3.2 Axes Are Perpendicular

When $\alpha_{12} = \pi/2$, (5.4) reduces to

$$a_{12} = 0, \tag{5.13}$$

and the two screws will be reciprocal only if they intersect. Thus, two screws that intersect at right angles will always be reciprocal no matter the value of the pitch of each individual screw.

5.3.3 Axes Are Parallel

When the lines of action of the wrench and the twist are parallel, as shown in Figure 5.3, $\alpha_{12} = 0$ and (5.4) reduces to

$$h_1 + h_2 = 0. \tag{5.14}$$

The wrench and twist will be reciprocal only if $h_1 = -h_2$. This is also the reciprocity condition when the lines of action of the wrench and twist are coaxial.

5.3.4 Body Constrained in Translation

Finally, consider the case whereby the wrench $f_2\$_2$ acts upon a body that is constrained to translate in a direction parallel to the vector $^0S^1$, as shown in Figure 5.4.

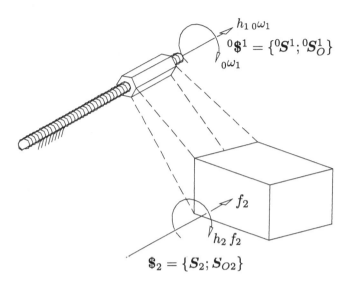

$h_1 \, _0\omega_1$

$^0\$^1 = \{^0S^1; \, ^0S_O^1\}$

$_0\omega_1$

f_2

$h_2 \, f_2$

$\$_2 = \{S_2; S_{O2}\}$

Figure 5.3 A wrench parallel to a twist

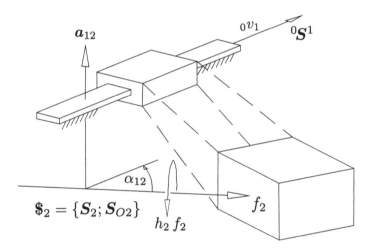

a_{12}

$_0v_1$ $^0S^1$

α_{12}

f_2

$\$_2 = \{S_2; S_{O2}\}$ $h_2 \, f_2$

Figure 5.4 A wrench applied to a body constrained in translation

The coordinates of the twist screw will be $_0v_1\{0; \, ^0S^1\}$, and the virtual work of the wrench about this twist will be simply

$$U = f_2 \, _0v_1 \cos\alpha_{12}. \tag{5.15}$$

Clearly, any applied wrench will cause the body to move unless the line of action of the wrench is perpendicular to the direction of the translation, that is, $\cos\alpha_{12} = 0$. When the line of action of the wrench is perpendicular to the direction of translation, no wrench, no matter what its pitch (such as a pure force when the pitch equals zero, or a pure couple when the pitch is infinite) will cause the body to move.

5.4   Reciprocal Screw Systems

5.4.1   A Single Screw: One Degree of Freedom

Consider a body connected to ground by a single screw $^0\$^1 = \{L_1, M_1, N_1; P_1, Q_1, R_1\}$, as shown in Figure 5.5. Any screw $\$_r = \{L_r, M_r, N_r; P_r, Q_r, R_r\}$ that is reciprocal to $^0\$^1$ must satisfy the relationship

$$L_1 P_r + M_1 Q_r + N_1 R_r + L_r P_1 + M_r Q_1 + N_r R_1 = 0. \qquad (5.16)$$

Since the screw coordinates are homogeneous, this equation may be written as

$$L_1 \frac{P_r}{L_r} + M_1 \frac{Q_r}{L_r} + N_1 \frac{R_r}{L_r} + P_1 + \frac{M_r}{L_r} Q_1 + \frac{N_r}{L_r} R_1 = 0, \qquad (5.17)$$

and it is apparent that there are five linearly independent screws that are reciprocal to $^0\$^1$. Any four of these ratios can be selected and the fifth determined by solving (5.17). Thus, there are ∞^4 screws in the reciprocal five system.

The coordinates of the screw $^0\$^1$ together with five linearly independent screws that are reciprocal to $^0\$^1$ can be written in terms of the coordinate system, as illustrated in Figure 5.5, as

$$
\begin{aligned}
^0\$^1 &= \{0, 0, 1; 0, 0, h_1\}, &\$_{r1} &= \{0, 0, 1; 0, 0, -h_1\}, \\
& & \$_{r2} &= \{1, 0, 0; h_2, 0, 0\}, \\
& & \$_{r3} &= \{1, 0, 0; -h_2, 0, 0\}, &&(5.18) \\
& & \$_{r4} &= \{0, 1, 0; 0, h_3, 0\}, \\
& & \$_{r5} &= \{0, 1, 0; 0, -h_3, 0\}.
\end{aligned}
$$

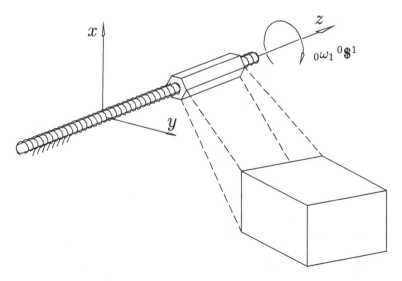

Figure 5.5 A body connected to ground by a single screw

The first screw reciprocal to $^0\$^1$ is a wrench, which is coaxial with $^0\$^1$ but with equal and opposite pitch. The second, third, fourth, and fifth screws reciprocal to $^0\$^1$ are wrenches coaxial with the x and y axes, and with pitches $\pm h_2$ and $\pm h_3$. It is interesting to note that each of the six screws tabulated is reciprocal with every other one of the tabulated screws, and such screws were called *co-reciprocal* by Ball.

Further, it is important to recognize that the five linearly independent reciprocal wrenches characterize the nature of the constraints acting upon the body. If the physical screw $^0\$^1$ was removed and these five wrenches were applied to the body, then the only motion possible would be a screwing motion about the axis of $^0\$^1$. Any other five linearly independent screws that are linear combinations of the above five reciprocal screws would, of course, provide the same constraints and permit the same relative motion of the body.

5.4.2 Two Screws: Two Degrees of Freedom

Figure 5.6 shows a rigid body that is connected to ground by a pair of serially connected screws $^0\$^1 = \{L_1, M_1, N_1; P_1, Q_1, R_1\}$ and $^1\$^2 = \{L_2, M_2, N_2; P_2, Q_2, R_2\}$. The body has two independent freedoms. The screw representing the instant motion of the body, $_0\omega_2 \,^0\$^2$, lies on a cylindroid, and it is a linear combination of the two screws representing the joint motions, i.e.,

$$_0\omega_2 \,^0\$^2 = {}_0\omega_1 \,^0\$^1 + {}_1\omega_2 \,^1\$^2. \tag{5.19}$$

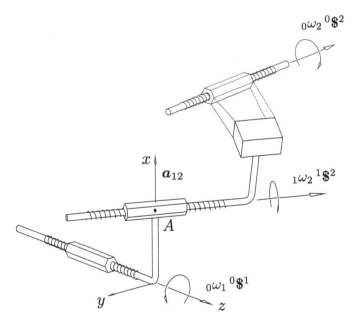

Figure 5.6 A body connected to ground by two screws

Any screw $\$_r = \{L_r, M_r, N_r; P_r, Q_r, R_r\}$ that is simultaneously reciprocal to $^0\$^1$ and $^1\$^2$ must satisfy the two equations

$$L_1 \frac{P_r}{L_r} + M_1 \frac{Q_r}{L_r} + N_1 \frac{R_r}{L_r} + P_1 + \frac{M_r}{L_r} Q_1 + \frac{N_r}{L_r} R_1 = 0, \tag{5.20}$$

$$L_2 \frac{P_r}{L_r} + M_2 \frac{Q_r}{L_r} + N_2 \frac{R_r}{L_r} + P_2 + \frac{M_r}{L_r} Q_2 + \frac{N_r}{L_r} R_2 = 0. \tag{5.21}$$

This pair of equations consist of four independent parameters, for example, $\frac{N_r}{L_r}$, $\frac{P_r}{L_r}$, $\frac{Q_r}{L_r}$, and $\frac{R_r}{L_r}$, and, therefore, there are four linearly independent screws that are simultaneously reciprocal to $^0\$^1$ and $^1\$^2$. Thus, there are four constraints on the body, which permit two independent freedoms. Further, any three of these ratios can be selected and the remaining two determined by the simultaneous solution of (5.20) and (5.21). Hence, there are ∞^3 screws in the reciprocal four system.

The coordinate system shown in Figure 5.6 was selected in order to simplify the expressions for the coordinates of the two motion screws. These are written in terms of this coordinate system as

$$^0\$^1 = \{0, 0, 1; 0, 0, h_1\}, \tag{5.22}$$
$$^1\$^2 = \{0, M_2, N_2; 0, Q_2, R_2\}.$$

Equations (5.20) and (5.21) reduce to

$$R_r + h_1 N_r = 0, \tag{5.23}$$

$$M_2 Q_r + N_2 R_r + M_r Q_2 + N_r R_2 = 0. \tag{5.24}$$

By inspection, a pair of linearly independent screws that are reciprocal to $^0\$^1$ and $^1\$^2$ are wrenches of equal and opposite pitches acting on the x axis with coordinates $\{1, 0, 0; \pm h_r, 0, 0\}$. It is not possible to determine a further two linearly independent reciprocal screws by inspection, but an additional reciprocal screw can be determined by arbitrarily selecting values for two of the reciprocal screw coordinates, say M_r and N_r, and then using (5.23) and (5.24) to solve for the remaining two screw coordinates, Q_r and R_r. This procedure can be repeated to yield a fourth reciprocal screw. It should be noted that the four reciprocal screws should be checked to ensure that they are linearly independent, and, if not, one of the screws can be replaced by a new reciprocal screw that is calculated by selecting different values for M_r and N_r.

A special case occurs if the two screws $^0\$^1$ and $^1\$^2$ have the same pitch, i.e., $h_1 = h_2 = h$. The coordinates of the two screws can be written as

$$^0\$^1 = \{^0S^1; h\,^0S^1\}, \tag{5.25}$$

$$^1\$^2 = \{^1S^2; a_{12}a_{12} \times \,^1S^2 + h\,^1S^2\}. \tag{5.26}$$

Expressing these screws in terms of the coordinate system shown in Figure 5.6 yields

$$^0\$^1 = \{0, 0, 1; 0, 0, h\}, \tag{5.27}$$

$$^1\$^2 = \{0, M_2, N_2; 0, -a_{12}N_2 + hM_2, a_{12}M_2 + hN_2\}. \tag{5.28}$$

For this case of equal pitch, a third screw that is reciprocal to $^0\$^1$ and $^1\$^2$ is a wrench with pitch $-h$, acting upon an axis that passes through the origin and that is parallel to $^1\$^2$. The coordinates of this screw can be written as

$$\$_{r3} = \{^1S^2; -h\,^1S^2\} = \{0, M_2, N_2; 0, -hM_2, -hN_2\}. \tag{5.29}$$

A fourth screw that is reciprocal to $^0\$^1$ and $^1\$^2$ is a wrench with pitch $-h$, acting upon an axis that passes through point A (see Figure 5.6) and that is parallel to $^0\$^1$. The coordinates of this screw can be written as

$$\$_{r4} = \{^0S^1; a_{12}\mathbf{a}_{12} \times {}^0S^1 - h\,^0S^1\} = \{0, 0, 1; 0, -a_{12}, -h\}. \tag{5.30}$$

Any other screw in this reciprocal screw system can be expressed as a linear combination of the above four screws.

5.4.3   Three Screws: Three Degrees of Freedom

Figure 5.7 shows a rigid body connected to ground by three serially connected screws. The screw $_0\omega_3\,^0S^3$ representing the instant motion of the body is a linear combination of the three screws representing the joint motions,

$$_0\omega_3\,^0\$^3 = {}_0\omega_1\,^0\$^1 + {}_1\omega_2\,^1\$^2 + {}_2\omega_3\,^2\$^3. \tag{5.31}$$

Any screw $\$_r = \{L_r, M_r, N_r; P_r, Q_r, R_r\}$ that is simultaneously reciprocal to $^0\$^1 = \{L_1, M_1, N_1; P_1, Q_1, R_1\}$, $^1\$^2 = \{L_2, M_2, N_2; P_2, Q_2, R_2\}$, and $^2\$^3 = \{L_3, M_3, N_3; P_3, Q_3, R_3\}$ must satisfy the three equations

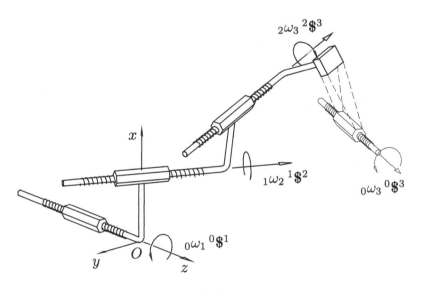

Figure 5.7 A body connected to ground by three screws

$$L_1 \frac{P_r}{L_r} + M_1 \frac{Q_r}{L_r} + N_1 \frac{R_r}{L_r} + P_1 + \frac{M_r}{L_r} Q_1 + \frac{N_r}{L_r} R_1 = 0, \tag{5.32}$$

$$L_2 \frac{P_r}{L_r} + M_2 \frac{Q_r}{L_r} + N_2 \frac{R_r}{L_r} + P_2 + \frac{M_r}{L_r} Q_2 + \frac{N_r}{L_r} R_2 = 0, \tag{5.33}$$

$$L_3 \frac{P_r}{L_r} + M_3 \frac{Q_r}{L_r} + N_3 \frac{R_r}{L_r} + P_3 + \frac{M_r}{L_r} Q_3 + \frac{N_r}{L_r} R_3 = 0. \tag{5.34}$$

These three equations consist of three independent parameters, for example, $\frac{P_r}{L_r}$, $\frac{Q_r}{L_r}$, and $\frac{R_r}{L_r}$, and, therefore, there are three linearly independent screws that are simultaneously reciprocal to $^0\$^1$, $^1\$^2$, and $^2\$^3$. Thus, there are three constraints that permit three independent freedoms. Further, any two of these three ratios can be selected and the remaining three determined by the simultaneous solution of (5.32), (5.33), and (5.34). Hence, there are ∞^2 screws in the reciprocal three system.

It is not possible to determine by inspection a wrench that is reciprocal to the three screws, but a screw reciprocal to the three motion screws can be determined by arbitrarily selecting values for three of the reciprocal screw coordinates, say L_r, M_r, and N_r, and then using (5.32), (5.33), and (5.34) to solve for the remaining three screw coordinates, P_r, Q_r, and R_r. This procedure can be repeated two more times to yield a total of three reciprocal screws. It should be noted that the three reciprocal screws should be checked to ensure that they are linearly independent, and, if not, one of the screws can be replaced by a new reciprocal screw that is calculated by selecting different values for L_r, M_r, and N_r.

When the pitches of the three screws are zero, $h_1 = h_2 = h_3 = 0$, the chain can be modeled by three serially connected revolute pairs, and a screw $\$_r$ that is simultaneously reciprocal to the lines $^0\$^1$, $^1\$^2$, and $^2\$^3$ can be determined geometrically. Figure 5.8 shows two planes. One is defined by a point A and the line $^0\$^1$, where A

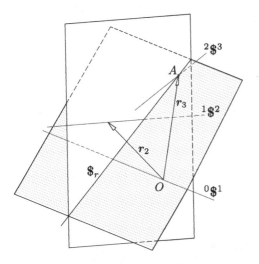

Figure 5.8 Screw reciprocal to three system ($h_1 = h_2 = h_3 = 0$)

is any point on the third line $^2\$^3$. The other is defined by point A and the line $^1\$^2$. These two planes intersect in a line reciprocal to the three motion screws. Therefore, a wrench reciprocal to three rotation screws is a force, $f\{S_r; S_{or}\}$, where $S_r \cdot S_{or} = 0$. Vectors normal to the two planes are, respectively, $r_3 \times {}^0S^1$ and $(r_2 - r_3) \times {}^1S^2$, where r_2 is a vector from reference point O on line $^0S^1$ to any point on line $^1S^2$, and r_3 is a vector from point O to point A on line $^2\$^3$. The equations of the planes are, therefore,

$$r \cdot \left(r_3 \times {}^0S^1\right) = 0, \tag{5.35}$$

$$(r - r_3) \cdot \left[(r_2 - r_3) \times {}^1S^2\right] = 0. \tag{5.36}$$

Equation (5.36) can be rewritten as

$$r \cdot \left[(r_2 - r_3) \times {}^1S^2\right] - r_3 \cdot \left[(r_2 - r_3) \times {}^1S^2\right] = 0. \tag{5.37}$$

Substituting $r_2 \times {}^1S^2 = {}^1S_O^2$ and recognizing that $r_3 \cdot r_3 \times {}^1S^2 = 0$ gives

$$r \cdot \left[{}^1S_O^2 - r_3 \times {}^1S^2\right] - r_3 \cdot {}^1S_O^2 = 0. \tag{5.38}$$

A vector parallel to the line of intersection of the planes is

$$\begin{aligned} S_r &= \left(r_3 \times {}^0S^1\right) \times \left((r_2 - r_3) \times {}^1S^2\right) \\ &= \left(r_3 \times {}^0S^1\right) \times \left({}^1S_O^2 - r_3 \times {}^1S^2\right). \end{aligned} \tag{5.39}$$

From (1.81), the equation of the line that is formed by the intersection of the two planes may be written as

$$r \times S_r = \left(r_3 \cdot {}^1S_O^2\right)\left[r_3 \times {}^0S^1\right]. \tag{5.40}$$

Equation (5.40) is a line through point A on $^2\$^3$ that intersects the lines $^0\$^1$ and $^1\$^2$. The three lines $^0\$^1$, $^1\$^2$, and $^2\$^3$ lie on a hyperboloid and, thus, as point A moves on $^2\$^3$ the line $\$_r$ will generate a hyperboloid that is complimentary to the hyperboloid defined by $^0\$^1$, $^1\$^2$, and $^2\$^3$. Therefore, it can be concluded that all the screws that are reciprocal to $^0\$^1$, $^1\$^2$, and $^2\$^3$ must lie on this complimentary hyperboloid. The same result is obtained when the three screws have the same pitch, $h_1 = h_2 = h_3 = h$, and the screws reciprocal to the three screws have pitch $\pm h$, as illustrated in Figure 5.9. It follows that screws of constant pitch in a three system lie on a series of concentric hyperboloids. When the coordinate system is chosen at the center of the hyperboloids, then any screw in the three system can be expressed as a linear combination of the three screws

$$\{1, 0, 0; h_x, 0, 0\}, \tag{5.41}$$
$$\{0, 1, 0; 0, h_y, 0\},$$
$$\{0, 0, 1; 0, 0, h_z\},$$

which are called the three principal screws and which themselves are *co-reciprocal*. The reciprocal three system of screws are wrenches, which can be expressed as a linear combination of the three screws

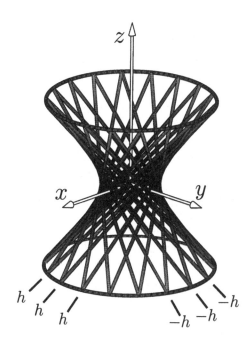

Figure 5.9 Hyperboloid of reciprocal screws ($h_1 = h_2 = h_3 = h$)

$$\{1, 0, 0; -h_x, 0, 0\}, \tag{5.42}$$
$$\{0, 1, 0; 0, -h_y, 0\},$$
$$\{0, 0, 1; 0, 0, -h_z\}.$$

5.4.4 Four Screws: Four Degrees of Freedom

The screw $_0\omega_4\,{}^0S^4$, which represents the instant motion of a body connected to ground by four serially connected screws, is a linear combination of the four screws representing the joint motions, i.e.,

$$_0\omega_4\,{}^0\$^4 = {}_0\omega_1\,{}^0\$^1 + {}_1\omega_2\,{}^1\$^2 + {}_2\omega_3\,{}^2\$^3 + {}_3\omega_4\,{}^3\$^4. \tag{5.43}$$

Any screw $\$_r = \{L_r, M_r, N_r; P_r, Q_r, R_r\}$ that is simultaneously reciprocal to ${}^0\$^1 = \{L_1, M_1, N_1; P_1, Q_1, R_1\}$, ${}^1\$^2 = \{L_2, M_2, N_2; P_2, Q_2, R_2\}$, ${}^2\$^3 = \{L_3, M_3, N_3; P_3, Q_3, R_3\}$, and ${}^3\$^4 = \{L_4, M_4, N_4; P_4, Q_4, R_4\}$ must satisfy the four equations

$$L_1 \frac{P_r}{L_r} + M_1 \frac{Q_r}{L_r} + N_1 \frac{R_r}{L_r} + P_1 + \frac{M_r}{L_r} Q_1 + \frac{N_r}{L_r} R_1 = 0, \tag{5.44}$$

$$L_2 \frac{P_r}{L_r} + M_2 \frac{Q_r}{L_r} + N_2 \frac{R_r}{L_r} + P_2 + \frac{M_r}{L_r} Q_2 + \frac{N_r}{L_r} R_2 = 0, \tag{5.45}$$

$$L_3 \frac{P_r}{L_r} + M_3 \frac{Q_r}{L_r} + N_3 \frac{R_r}{L_r} + P_3 + \frac{M_r}{L_r} Q_3 + \frac{N_r}{L_r} R_3 = 0, \tag{5.46}$$

$$L_4 \frac{P_r}{L_r} + M_4 \frac{Q_r}{L_r} + N_4 \frac{R_r}{L_r} + P_4 + \frac{M_r}{L_r} Q_4 + \frac{N_r}{L_r} R_4 = 0. \qquad (5.47)$$

These four equations consist of two independent parameters, for example, $\frac{Q_r}{L_r}$ and $\frac{R_r}{L_r}$, and, therefore, there are two linearly independent screws that are simultaneously reciprocal to $^0\$^1$, $^1\$^2$, $^2\$^3$, and $^3\$^4$. Thus, there are two constraints that permit four independent freedoms. Further, any one of these ratios can be selected and the remaining four determined by the simultaneous solution of (5.44) through (5.47). Hence, there are ∞^1 screws in the reciprocal two system, and all of these screws lie on a cylindroid.

5.4.5   Five Screws: Five Degrees of Freedom

The screw $_0\omega_5\, ^0S^5$, which represents the instant motion of a body connected to ground by five serially connected screws, is a linear combination of the five screws representing the joint motions, i.e.,

$$_0\omega_5\, ^0\$^5 = {_0\omega_1}\, ^0\$^1 + {_1\omega_2}\, ^1\$^2 + {_2\omega_3}\, ^2\$^3 + {_3\omega_4}\, ^3\$^4 + {_4\omega_5}\, ^4\$^5. \qquad (5.48)$$

Any screw $\$_r = \{L_r, M_r, N_r; P_r, Q_r, R_r\}$ that is simultaneously reciprocal to $^0\$^1 = \{L_1, M_1, N_1; P_1, Q_1, R_1\}$, $^1\$^2 = \{L_2, M_2, N_2; P_2, Q_2, R_2\}$, $^2\$^3 = \{L_3, M_3, N_3; P_3, Q_3, R_3\}$, $^3\$^4 = \{L_4, M_4, N_4; P_4, Q_4, R_4\}$, and $^4\$^5 = \{L_5, M_5, N_5; P_5, Q_5, R_5\}$ must satisfy the five equations

$$L_1 \frac{P_r}{L_r} + M_1 \frac{Q_r}{L_r} + N_1 \frac{R_r}{L_r} + P_1 + \frac{M_r}{L_r} Q_1 + \frac{N_r}{L_r} R_1 = 0, \qquad (5.49)$$

$$L_2 \frac{P_r}{L_r} + M_2 \frac{Q_r}{L_r} + N_2 \frac{R_r}{L_r} + P_2 + \frac{M_r}{L_r} Q_2 + \frac{N_r}{L_r} R_2 = 0, \qquad (5.50)$$

$$L_3 \frac{P_r}{L_r} + M_3 \frac{Q_r}{L_r} + N_3 \frac{R_r}{L_r} + P_3 + \frac{M_r}{L_r} Q_3 + \frac{N_r}{L_r} R_3 = 0, \qquad (5.51)$$

$$L_4 \frac{P_r}{L_r} + M_4 \frac{Q_r}{L_r} + N_4 \frac{R_r}{L_r} + P_4 + \frac{M_r}{L_r} Q_4 + \frac{N_r}{L_r} R_4 = 0, \qquad (5.52)$$

$$L_5 \frac{P_r}{L_r} + M_5 \frac{Q_r}{L_r} + N_5 \frac{R_r}{L_r} + P_5 + \frac{M_r}{L_r} Q_5 + \frac{N_r}{L_r} R_5 = 0. \qquad (5.53)$$

These five equations consist of one independent parameter, for example, $\frac{R_r}{L_r}$, and, therefore, there is one linearly independent screw that is simultaneously reciprocal to $^0\$^1$, $^1\$^2$, $^2\$^3$, $^3\$^4$, and $^4\$^5$. Thus, there is one constraint that permits five independent freedoms. Further, all five ratios can be determined by solving the set of equations (5.49) through (5.53). Hence, there is one screw that is reciprocal to the five motion screws.

5.4.6   Six Screws: Six Degrees of Freedom

The screw $_0\omega_6\, ^0S^6$, which represents the instant motion of a body connected to ground by six serially connected screws, is a linear combination of the six screws representing the joint motions, i.e.,

$$_0\omega_6 \, ^0\$^6 = _0\omega_1 \, ^0\$^1 + _1\omega_2 \, ^1\$^2 + _2\omega_3 \, ^2\$^3 + _3\omega_4 \, ^3\$^4 + _4\omega_5 \, ^4\$^5 + _5\omega_6 \, ^5\$^6. \qquad (5.54)$$

In general, there is no screw that is simultaneously reciprocal to the six motion screws. The condition for reciprocity, however, can be written in matrix form as

$$\begin{bmatrix} P_1 & Q_1 & R_1 & L_1 & M_1 & N_1 \\ P_2 & Q_2 & R_2 & L_2 & M_2 & N_2 \\ P_3 & Q_3 & R_3 & L_3 & M_3 & N_3 \\ P_4 & Q_4 & R_4 & L_4 & M_4 & N_4 \\ P_5 & Q_5 & R_5 & L_5 & M_5 & N_5 \\ P_6 & Q_6 & R_6 & L_6 & M_6 & N_6 \end{bmatrix} \begin{bmatrix} L_r \\ M_r \\ N_r \\ P_r \\ Q_r \\ R_r \end{bmatrix} = \begin{bmatrix} 0 \\ 0 \\ 0 \\ 0 \\ 0 \\ 0 \end{bmatrix}, \qquad (5.55)$$

where the components L_i through R_i, $i = 1\ldots 6$, represent the components of the six motion screws, and L_r through R_r represent the components of the reciprocal screw. It is apparent that the only solution to these six homogeneous equations is the trivial solution $L_r = M_r = N_r = P_r = Q_r = R_r = 0$, unless the six motion screws are linearly dependent. This condition is discussed in Chapter 6.

5.4.7   Example Problem 1

A one system consisting of a single revolute joint is shown in Figure 5.10. The coordinates of the screw $^0\$^1$ are written as

$$^0\$^1 = \{0, 0, 1; 0, 0, 0\}. \qquad (5.56)$$

The objective is to determine by inspection the coordinates of five linearly independent screws that are reciprocal to the one system. By inspection, the reciprocal screws can be selected as forces along the three coordinate axes and moments about the x and y axes, as shown in Figure 5.10. The coordinates of these screws are written as

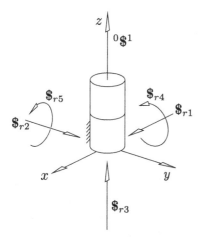

Figure 5.10 Five screws reciprocal to a one system

$$\$_{r1} = \{1, 0, 0; 0, 0, 0\},\qquad(5.57)$$
$$\$_{r2} = \{0, 1, 0; 0, 0, 0\},$$
$$\$_{r3} = \{0, 0, 1; 0, 0, 0\},$$
$$\$_{r4} = \{0, 0, 0; 1, 0, 0\},$$
$$\$_{r5} = \{0, 0, 0; 0, 1, 0\}.$$

5.4.8   Example Problem 2

A two system consisting of two revolute joints is shown at a particular instant in Figure 5.11. The coordinates of the screws $^0\$^1$ and $^1\$^2$ are written as

$$^0\$^1 = \{0, 0, 1; 0, 0, 0\},\qquad(5.58)$$
$$^1\$^2 = \{1, 0, 0; 0, Q_2, 0\}.$$

The objective is to determine by inspection the coordinates of four linearly independent screws that are reciprocal to the two system. By inspection, the reciprocal screws can be selected as forces along the x and z axes, a moment about the y axis, and a force parallel to the y axis that passes through the line of action of $^1\$^2$, as shown in Figure 5.11. The coordinates of these screws are written as

$$\$_{r1} = \{1, 0, 0; 0, 0, 0\},\qquad(5.59)$$
$$\$_{r2} = \{0, 0, 1; 0, 0, 0\},$$
$$\$_{r3} = \{0, 0, 0; 0, 1, 0\},$$
$$\$_{r4} = \{0, 1, 0; -Q_2, 0, 0\}.$$

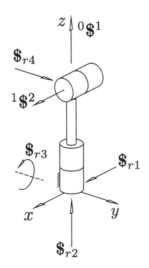

Figure 5.11 Four screws reciprocal to a two system

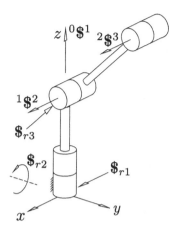

Figure 5.12 Three screws reciprocal to a three system

5.4.9 Example Problem 3

A three system consisting of three revolute joints is shown at a particular instant in Figure 5.12. The coordinates of the screws $^0\$^1$, $^1\$^2$, and $^2\$^3$ are written as

$$^0\$^1 = \{0, 0, 1; 0, 0, 0\}, \tag{5.60}$$
$$^1\$^2 = \{1, 0, 0; 0, Q_2, 0\},$$
$$^2\$^3 = \{1, 0, 0; 0, Q_3, R_3\}.$$

The objective is to determine by inspection the coordinates of three linearly independent screws that are reciprocal to the three system. By inspection, the reciprocal screws can be selected as a force along the x axis, a moment about the y axis, and a force that intersects all three motion screws, as shown in Figure 5.12. The coordinates of these screws are written as

$$\$_{r1} = \{1, 0, 0; 0, 0, 0\}, \tag{5.61}$$
$$\$_{r2} = \{0, 0, 0; 0, 1, 0\},$$
$$\$_{r3} = \left\{0, 1, \frac{Q_2 - Q_3}{R_3}; -Q_2, 0, 0\right\}.$$

5.4.10 Example Problem 4

A four system consisting of four revolute joints is shown at a particular instant in Figure 5.13. The coordinates of the screws $^0\$^1$, $^1\$^2$, $^2\$^3$, and $^3\$^4$ are written as

$$^0\$^1 = \{0, 0, 1; 0, 0, 0\}, \tag{5.62}$$
$$^1\$^2 = \{1, 0, 0; 0, Q_2, 0\},$$
$$^2\$^3 = \{1, 0, 0; 0, Q_3, R_3\},$$
$$^3\$^4 = \{1, 0, 0; 0, Q_4, R_4\}.$$

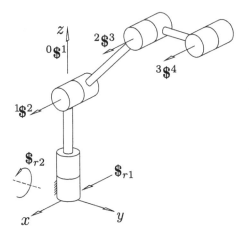

Figure 5.13 Two screws reciprocal to a four system

The objective is to determine by inspection the coordinates of two linearly independent screws that are reciprocal to the four system. By inspection, the reciprocal screws can be selected as a force along the x axis and a moment about the y axis, as shown in Figure 5.13. The coordinates of these screws are written as

$$\$_{r1} = \{1, 0, 0; 0, 0, 0\}, \tag{5.63}$$
$$\$_{r2} = \{0, 0, 0; 0, 1, 0\}.$$

5.4.11   Example Problem 5

A five system consisting of five revolute joints is shown at a particular instant in Figure 5.14. The coordinates of the screws $^0\$^1$, $^1\$^2$, $^2\$^3$, $^3\$^4$, and $^4\$^5$ are written as

$$^0\$^1 = \{0, 0, 1; 0, 0, 0\}, \tag{5.64}$$
$$^1\$^2 = \{1, 0, 0; 0, Q_2, 0\},$$
$$^2\$^3 = \{1, 0, 0; 0, Q_3, R_3\},$$
$$^3\$^4 = \{1, 0, 0; 0, Q_4, R_4\},$$
$$^4\$^5 = \{0, M_5, N_5; P_5, Q_5, R_5\}.$$

The objective is to determine by inspection the coordinates of one screw that is reciprocal to the five system. By inspection, the reciprocal screw can be selected as a force parallel to the x axis that intersects $^0\$^1$ (the z axis) and the fifth screw $^4\$^5$, as shown in Figure 5.14. The coordinates of this screw are written as

$$\$_{r1} = \left\{ 1, 0, 0; 0, \frac{-P_5}{M_5}, 0 \right\}. \tag{5.65}$$

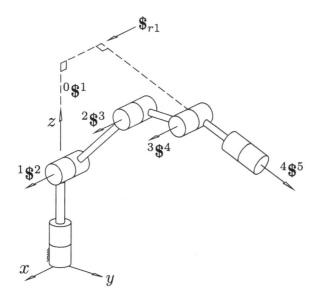

Figure 5.14 One screw reciprocal to a five system

5.5   Velocity Analysis of Serial Manipulators Using Reciprocal Screws

It was shown in Chapter 4 that the instantaneous velocity state of the end effector of a
six degree of freedom serial manipulator can be obtained from the screw equation

$$_0\omega_6 \, ^0\$^6 = {}_0\omega_1 \, ^0\$^1 + {}_1\omega_2 \, ^1\$^2 + {}_2\omega_3 \, ^2\$^3 + {}_3\omega_4 \, ^3\$^4 + {}_4\omega_5 \, ^4\$^5 + {}_5\omega_6 \, ^5\$^6. \quad (5.66)$$

This equation can be written in matrix format as

$$_0\omega_6 \, ^0\$^6 = \boldsymbol{J}\,\boldsymbol{\omega}, \quad (5.67)$$

where $\boldsymbol{J}$, named the Jacobian matrix, is the 6×6 matrix formed from the unitized
screw coordinates as

$$\boldsymbol{J} = \begin{bmatrix} ^0\$^1 & ^1\$^2 & ^2\$^3 & ^3\$^4 & ^4\$^5 & ^5\$^6 \end{bmatrix} \quad (5.68)$$

$$= \begin{bmatrix} ^0S^1 & ^1S^2 & ^2S^3 & ^3S^4 & ^4S^5 & ^5S^6 \\ ^0S_O^1 & ^1S_O^2 & ^2S_O^3 & ^3S_O^4 & ^4S_O^5 & ^5S_O^6 \end{bmatrix}$$

and $\boldsymbol{\omega}$ is the vector of joint velocities given by

$$\boldsymbol{\omega} = \begin{bmatrix} _0\omega_1 \\ _1\omega_2 \\ _2\omega_3 \\ _3\omega_4 \\ _4\omega_5 \\ _5\omega_6 \end{bmatrix}. \quad (5.69)$$

For the reverse velocity analysis problem, the desired velocity state of the end effector is specified together with the screws $^0\$^1$ through $^5\$^6$. The objective is to determine the individual joint velocities $_0\omega_1$ through $_5\omega_6$ that will cause the end effector to move as desired. In Chapter 4, it was shown how the joint velocities can be determined by inverting the Jacobian matrix to yield

$$\omega = J^{-1}{}_0\omega_6\, ^0\$^6. \tag{5.70}$$

Here, an alternate solution for ω is presented that uses reciprocal screws.

First, a single screw that is reciprocal to the first five screws $^0\$^1$, $^1\$^2$, $^2\$^3$, $^3\$^4$, and $^4\$^5$ can be determined by selecting a value for L_r and then solving the equations (5.49) through (5.53) for the remaining components of the reciprocal screw. This reciprocal screw will be named $\$_{r5}$. Forming the reciprocal product of the left and right sides of (5.66) with $\$_{r5}$ gives

$$_0\omega_6\, ^0\$^6 \circ \$_{r5} = {}_5\omega_6\, ^5\$^6 \circ \$_{r5} \tag{5.71}$$

and, therefore,

$$_5\omega_6 = \frac{^0\$^6 \circ \$_{r5}}{^5\$^6 \circ \$_{r5}}{}_0\omega_6. \tag{5.72}$$

Following this, a screw that is reciprocal to the first four screws in the chain, i.e., $^0\$^1$, $^1\$^2$, $^2\$^3$, and $^3\$^4$, can be determined by selecting values for L_r and M_r and then obtaining the other components by solving (5.44) through (5.47). This screw will be called $\$_{r4}$, and forming the reciprocal product of this screw with the left and right sides of (5.66) gives

$$_0\omega_6\, ^0\$^6 \circ \$_{r4} = {}_4\omega_5\, ^4\$^5 \circ \$_{r4} + {}_5\omega_6\, ^5\$^6 \circ \$_{r4}. \tag{5.73}$$

Solving for $_4\omega_5$ gives

$$_4\omega_5 = \frac{{}_0\omega_6\left(^0\$^6 \circ \$_{r4}\right) - {}_5\omega_6\left(^5\$^6 \circ \$_{r4}\right)}{^4\$^5 \circ \$_{r4}}. \tag{5.74}$$

Similarly, screws $\$_{r3}$, $\$_{r2}$, and $\$_{r1}$ can be found, which are reciprocal to the sub-systems $\{^0\$^1, {}^1\$^2, {}^2\$^3\}$, $\{^0\$^1, {}^1\$^2\}$, and $\{^0\$^1\}$, respectively, and expressions for $_3\omega_4$, $_2\omega_3$, and $_1\omega_2$ can be obtained as

$$_3\omega_4 = \frac{{}_0\omega_6\left(^0\$^6 \circ \$_{r3}\right) - {}_4\omega_5\left(^4\$^5 \circ \$_{r3}\right) - {}_5\omega_6\left(^5\$^6 \circ \$_{r3}\right)}{^3\$^4 \circ \$_{r3}}, \tag{5.75}$$

$$_2\omega_3 = \frac{{}_0\omega_6\left(^0\$^6 \circ \$_{r2}\right) - {}_3\omega_4\left(^3\$^4 \circ \$_{r2}\right) - {}_4\omega_5\left(^4\$^5 \circ \$_{r2}\right) - {}_5\omega_6\left(^5\$^6 \circ \$_{r2}\right)}{^2\$^3 \circ \$_{r2}},$$
$$\tag{5.76}$$

and

$$_1\omega_2 = \frac{1}{^1\$^2 \circ \$_{r1}}\Big[{}_0\omega_6\left(^0\$^6 \circ \$_{r1}\right) - {}_2\omega_3\left(^2\$^3 \circ \$_{r1}\right) - {}_3\omega_4\left(^3\$^4 \circ \$_{r1}\right)$$
$$- {}_4\omega_5\left(^4\$^5 \circ \$_{r1}\right) - {}_5\omega_6\left(^5\$^6 \circ \$_{r1}\right)\Big]. \tag{5.77}$$

The first joint velocity, $_0\omega_1$, may then be determined from (5.54).

Clearly, for robots with general dimensions, much computation is required to determine the reciprocal screws $\$_{r3}$, $\$_{r4}$, and $\$_{r5}$. However, appropriate reciprocal screws $\$_{r1}$ through $\$_{r5}$ can be determined by inspection for industrial robots with special geometry, such as three intersecting or three parallel revolute joints. Thus, in these cases the solution presented here may well be preferred to inverting the Jacobian matrix as the number of computations (multiplications and additions) is reduced. It must be noted, however, that the sequential solution of the joint velocities may lead to cumulative errors.

5.5.1   Example Problem

In Section 4.7.1, the reverse velocity analysis was conducted for a Puma manipulator. The unitized coordinates for the six joint twists were calculated for the given manipulator configuration as

$$^0\$^1 = \{0, 0, 1; 0, 0, 0\},$$

$$^1\$^2 = \{-0.7071, 0.707, 0; 0, 0, 0\},$$

$$^2\$^3 = \{-0.7071, 0.7071, 0; -6.0104, -6.0104, 14.7224\},$$

$$^3\$^4 = \{0.7071, 0.7071, 0; -6.5761, 6.5761, -5.900\},$$

$$^4\$^5 = \{-0.5, 0.5, 0.7071; 14.1612, -17.5612, 22.4311\},$$

$$^5\$^6 = \{0.0795, -0.7866, 0.6124; 23.6061, -10.4425, -16.4759\},$$

where the first three elements of each screw are dimensionless, and the last three elements of each screw have units of inches. The desired velocity state for the last link of the manipulator was given as

$$\{^0\omega^6; {}^0v_0^6\} = \{0.20, -0.12, 0.10; 10.6913, 4.7221, -10.9160\},$$

where the first three terms have units of rad/sec and the last three have units of in/sec. It is desired to obtain the angular velocity of each joint axis that will cause the last link of the manipulator to move as specified.

The first step is to determine a screw $\$_{r5}$ that is simultaneously reciprocal to the first five screws of the manipulator. This can be accomplished by selecting a value for $L_r = 1$ and then solving the equations (5.49) through (5.53) for the remaining components of the reciprocal screw. The calculated values for $\$_{r5}$ are

$$\$_{r5} = \{1, 2.7748, 1.5411; -1.8236, -1.8236, 0\},$$

where the first three terms are dimensionless and the last three have units of inches. The angular velocity $_5\omega_6$ can now be solved from (5.72) as

$$_5\omega_6 = \frac{^0\$^6 \circ \$_{r5}}{^5\$^6 \circ \$_{r5}} \, _0\omega_6 = -0.2316 \text{ rad/sec.}$$

The next step is to determine a screw $\$_{r4}$ that is simultaneously reciprocal to the first four screws of the manipulator. This can be accomplished by selecting the values $L_r = 1$, $M_r = 1$ and then solving the equations (5.44) through (5.47) for the remaining components of the reciprocal screw. The calculated values for $\$_{r4}$ are

$$\$_{r4} = \{1, 1, 0.8165; 3.4064, 3.4064, 0\},$$

where the first three terms are dimensionless and the last three have units of inches. The angular velocity $_4\omega_5$ can now be solved from (5.74) as

$$_4\omega_5 = 0.4122 \text{ rad/sec.}$$

The screw $\$_{r3}$, which is simultaneously reciprocal to the first three screws of the manipulator, is determined next. From inspection, a pure force that passes through the origin whose direction is parallel to the second and third joint axes will be reciprocal to the first three motion screws. Thus,

$$\$_{r3} = \{-0.707, 0.707, 0; 0, 0, 0\}.$$

The angular velocity $_4\omega_5$ can now be solved from (5.75) as

$$_3\omega_4 = -0.0593 \text{ rad/sec.}$$

The screw $\$_{r2}$, which is simultaneously reciprocal to the first two screws of the manipulator, is determined next. From inspection, any pure force that passes through the origin will be reciprocal to the first two motion screws. Thus, the screw $\$_{r2}$ is selected as

$$\$_{r2} = \{0, 1, 1; 0, 0, 0\}.$$

The angular velocity $_2\omega_3$ can now be solved from (5.76) as

$$_2\omega_3 = -1.6525 \text{ rad/sec.}$$

The next step is to determine a screw $\$_{r1}$ that is reciprocal to only the first motion screw. By inspection, the coordinates of this screw were selected to be

$$\$_{r1} = \{1, 0, 0; 1, 0, 0\}.$$

The angular velocity $_1\omega_2$ can now be solved from (5.77) as

$$_1\omega_2 = 0.9929 \text{ rad/sec.}$$

The last angular velocity to be determined is $_0\omega_1$. This term can be obtained from the third of the six scalar equations represented by (5.66) as

$$_0\omega_1 = -0.0497 \text{ rad/sec.}$$

The results obtained here agree with those obtained in Section 4.7.1.

5.6   Forward and Reverse Velocity Analysis for Parallel Mechanisms

Figure 5.15 shows an in-parallel mechanism where the top platform is connected to the base platform by six S-P-H kinematic chains. It is assumed that the position analysis has been completed and, thus, the coordinates of the end points of each leg connector and thereby the Plücker coordinates of the lines along the six connectors, i.e., $\$_1$ through $\$_6$, are known.

5.6.1   Reverse Velocity Analysis

For the reverse velocity problem, the desired motion of the top platform has been given as $\left\{ {}^b\omega^t; \; {}^b v_0^t \right\}$, where the letters b and t refer to the base and top platform bodies, respectively. It is necessary to determine the linear velocities along each prismatic joint of each connector so that the top platform will move as desired.

One solution to this problem is straightforward. Since the velocity state of the top platform is given, the velocity of the top point for each leg connector can obviously be obtained. The projection of this velocity onto the unit vector S_i will yield the slider joint velocity for leg i.

A second solution is presented where the kinematic chain of each leg connector is analyzed. For a particular connector, the ball joint will be modeled by three intersecting revolute joints and the Hooke joint by two intersecting revolute joints. The velocity equation for each of the six legs can be written as

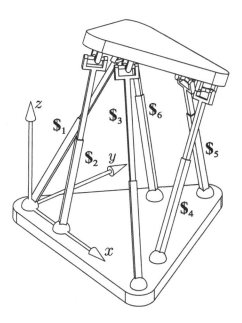

Figure 5.15 Parallel mechanism

$$\{^b\omega^t;{^b}v_0^t\} = (_b\omega_1)_i \, ^b\$_i^1 + (_1\omega_2)_i \, ^1\$_i^2 + (_2\omega_3)_i \, ^2\$_i^3 \qquad (5.78)$$
$$+ (_3v_4)_i \, ^3\$_i^4 + (_4\omega_5)_i \, ^4\$_i^5 + (_5\omega_t)_i \, ^5\$_i^t, \qquad i = 1 \ldots 6.$$

The screw coordinates for $^3\$_i^4$, associated with the prismatic joint may be written as $\{0; S_i\}$, where S_i is a unit vector from the base point of leg i to the top point of leg i. The other screw coordinates for the other joints in connector i can be obtained since the geometric parameters of the modeled ball joint and Hooke joint would be known. However, it will not be necessary to obtain all these line coordinates.

For leg i, a screw can be found that is reciprocal to all the screws in the leg except for the prismatic joint. This screw will be a pure force that passes through the center of the ball joint and the center of the Hooke joint. This reciprocal screw will be called $\$_{Ri}$, and its coordinates can be written as

$$\$_{Ri} = \{S_i; S_{OLi}\} = \{S_i; p_{bi} \times S_i\}, \qquad (5.79)$$

where p_{bi} is a vector from the origin of the reference system to the point at the center of the spherical joint for leg i. Evaluating the reciprocal product of equation (5.78) with $\$_{Ri}$ for leg i will give

$$\{^b\omega^t;{^b}v_0^t\} \circ \$_{Ri} = (_3v_4)_i \, ^3\$_i^4 \circ \$_{Ri} \qquad (5.80)$$

and substituting $^3\$_i^4 \circ \$_{Ri} = \{0; S_i\} \circ \{S_i; p_{bi} \times S_i\} = 1$ gives

$$(_3v_4)_i = \{^b\omega^t;{^b}v_0^t\} \circ \$_{Ri}. \qquad (5.81)$$

This equation can be simplified by writing the velocity of the prismatic joint along leg i as v_i to yield

$$v_i = \{^b\omega^t;{^b}v_0^t\} \circ \$_{Ri}, \quad i = 1 \ldots 6. \qquad (5.82)$$

5.6.2   Forward Velocity Analysis

For the forward velocity problem, the linear velocity of the prismatic joints of the six leg connectors will be known, and it is necessary to calculate the velocity state of the top platform with respect to the base. Again, it is assumed that the position problem has been completed and, therefore, the coordinates of the end points of each leg connector are known. Substituting (5.79) into (5.82) and expanding the reciprocal product gives

$$v_i = {^b}\omega^t \cdot S_{OLi} + {^b}v_0^t \cdot S_i, \quad i = 1 \ldots 6. \qquad (5.83)$$

These six equations can be written in matrix form as

$$v = J^T \begin{bmatrix} {^b}v_0^t \\ {^b}\omega^t \end{bmatrix}, \qquad (5.84)$$

where J is a 6×6 matrix formed from the Plücker coordinates of the lines along the leg connectors as

$$J = \begin{bmatrix} S_1 & S_2 & S_3 & S_4 & S_5 & S_6 \\ S_{OL1} & S_{OL2} & S_{OL3} & S_{OL4} & S_{OL5} & S_{OL6} \end{bmatrix} \tag{5.85}$$

and v is a length six vector written as

$$v = [v_1, v_2, v_3, v_4, v_5, v_6]^T. \tag{5.86}$$

Solving (5.84) for the velocity state of the top platform yields

$$\begin{bmatrix} {}^b v_0^t \\ {}^b \omega^t \end{bmatrix} = (J^T)^{-1} v. \tag{5.87}$$

5.6.3   Example Problem

At a particular instant, the coordinates of the center of each spherical joint and each Hooke joint of each leg connector of the parallel platform shown in Figure 5.15 are given in Table 5.1. At this instant, the desired velocity state of the top platform is given as

$$\{{}^b \omega^t; {}^b v_0^t\} = \{0, 0.5, 0.5; 11.2, 9.4, 3.7\},$$

where the first three terms have units of rad/sec and the last three have units of m/sec. The objective is to determine the linear velocities of the prismatic joints that will cause the top platform to move as specified.

The Plücker coordinates of the line along each leg connector can be determined as

$$S_i = \frac{p_{ti} - p_{bi}}{|p_{ti} - p_{bi}|},$$
$$S_{OLi} = p_{bi} \times S_i, \quad i = 1 \dots 6,$$

where p_{bi} and p_{ti} represent the coordinates of the base point and top point of leg connector i, respectively. The Plücker coordinates are evaluated as

Table 5.1. Coordinates of connector end points (units in m)

Connector #	Base Point	Top Point
1	$(0, 0, 0)$	$(1.31, 3.17, 6.11)$
2	$(1.74, 0, 0)$	$(3.94, 0, 6.47)$
3	$(10, 0, 0)$	$(4.23, 0.817, 6.54)$
4	$(9.13, 1.51, 0)$	$(5.61, 4.69, 6.90)$
5	$(5, 8.66, 0)$	$(4.77, 4.54, 6.75)$
6	$(4.13, 7.15, 0)$	$(0.760, 3.84, 6.03)$

$$\{S_1; S_{OL1}\} = \{0.187, 0.452, 0.872; 0, 0, 0\},$$
$$\{S_2; S_{OL2}\} = \{0.322, 0, 0.947; 0, -1.647, 0\},$$
$$\{S_3; S_{OL3}\} = \{-0.659, 0.093, 0.747; 0, -7.466, 0.933\},$$
$$\{S_4; S_{OL4}\} = \{-0.420, 0.380, 0.824; 1.244, -7.524, 4.102\},$$
$$\{S_5; S_{OL5}\} = \{-0.029, -0.521, 0.853; 7.389, -4.266, -2.352\},$$
$$\{S_6; S_{OL6}\} = \{-0.439, -0.432, 0.787; 5.629, -3.251, 1.361\},$$

where the first three components are dimensionless and the last three have units of meters. The linear velocities for each of the prismatic joints is then calculated from (5.83) as

$$v_1 = 9.573 \text{ m/sec},$$
$$v_2 = 6.285 \text{ m/sec},$$
$$v_3 = -7.005 \text{ m/sec},$$
$$v_4 = 0.200 \text{ m/sec},$$
$$v_5 = -5.373 \text{ m/sec},$$
$$v_6 = -7.022 \text{ m/sec}.$$

5.7   Problems

1. For the parallel platform shown in Figure 5.15, whose base and top point leg coordinates at this instant are given in Table 5.1, determine the velocity state of the top platform, $\{^b\omega^t; ^bv_0^t\}$, if the prismatic joint velocities at this instant are $v_1 = 2$ m/sec, $v_2 = -2$ m/sec, $v_3 = 1.5$ m/sec, $v_4 = 4.2$ m/sec, $v_5 = -1$ m/sec, and $v_6 = 0.5$ m/sec.

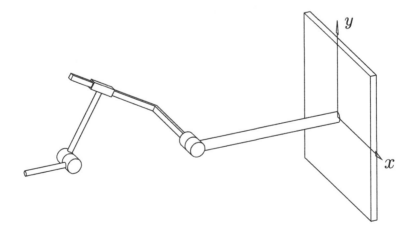

Figure 5.16 RPR manipulator

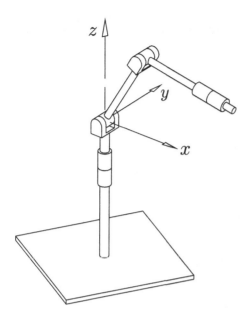

Figure 5.17 Four axis manipulator

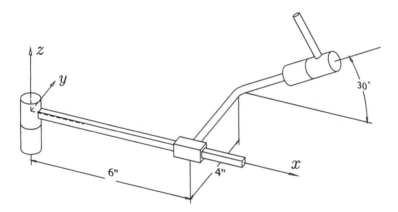

Figure 5.18 RPR manipulator

2. Repeat Problem 4.3 using the reciprocal screw approach to solve the reverse velocity problem for the Cincinnati Milacron T3-776 robot manipulator.
3. Identify three linearly independent screws that are simultaneously reciprocal to the three motion twists of the RPR manipulator shown in Figure 5.16.
4. Identify two linearly independent screws that are simultaneously reciprocal to the four motion twists of the manipulator shown in Figure 5.17.
5. Figure 5.18 shows a three degree of freedom robot manipulator comprised of a revolute, a prismatic, and another revolute joint. Determine the coordinates of three linearly independent wrenches that would produce no motion if applied to the end effector. Write the coordinates in terms of the coordinate system shown.

6   Singularity Analysis of Serial Chains

Nay, soaring still further into the empyrean,
he showed that all of the instantaneous motions of
every molecule in the universe were only a twist
about one screw chain, while all the forces of the
universe were but a wrench upon another.

Speaking of a committee member
Ball's "*A Dynamical Parable*" (1887)

6.1   Introduction

In general, it is possible to obtain the joint velocity parameters for a six degree of freedom serial robot manipulator to achieve any desired velocity state of the manipulator
end effector with respect to ground. However, this is not the case if the coordinates
of the lines (or screws) associated with each joint become linearly dependent. This
chapter addresses these concepts. All the configurations where the inverse velocity
problem fails are identified for several common manipulator geometries.

6.2   Singular Configurations

It was shown in Chapter 4 that the velocity state of the last link of a six degree of freedom manipulator, measured with respect to the grounded base, could be expressed as

$$\begin{bmatrix} {}^0\omega^6 \\ {}^0v_O^6 \end{bmatrix} = J\,\omega, \tag{6.1}$$

where J is a 6×6 matrix whose columns are the screws (with unitized direction) that
relate the instantaneous motion of body $i+1$ relative to body i, $i = 0\ldots5$, and ω is a
6×1 vector comprised of the magnitudes of the joint relative velocities. For the case
of a manipulator made up of six revolute joints,

$$\begin{aligned} J &= \begin{bmatrix} {}^0\$^1 & {}^1\$^2 & {}^2\$^3 & {}^3\$^4 & {}^4\$^5 & {}^5\$^6 \end{bmatrix} \\ &= \begin{bmatrix} {}^0S^1 & {}^1S^2 & {}^2S^3 & {}^3S^4 & {}^4S^5 & {}^5S^6 \\ {}^0S_{OL}^1 & {}^1S_{OL}^2 & {}^2S_{OL}^3 & {}^3S_{OL}^4 & {}^4S_{OL}^5 & {}^5S_{OL}^6 \end{bmatrix} \end{aligned} \tag{6.2}$$

and

$$\omega = \begin{bmatrix} _0\omega_1 \\ _1\omega_2 \\ _2\omega_3 \\ _3\omega_4 \\ _4\omega_5 \\ _5\omega_6 \end{bmatrix}. \tag{6.3}$$

If, say, body 4 were connected to body 3 by a prismatic joint whose direction of translation was defined by the unit vector $^3S^4$, then the matrix J and the vector ω could be written as

$$J = \begin{bmatrix} ^0S^1 & ^1S^2 & ^2S^3 & \mathbf{0} & ^4S^5 & ^5S^6 \\ ^0S_{OL}^1 & ^1S_{OL}^2 & ^2S_{OL}^3 & ^3S^4 & ^4S_{OL}^5 & ^5S_{OL}^6 \end{bmatrix}, \tag{6.4}$$

and

$$\omega = \begin{bmatrix} _0\omega_1 \\ _1\omega_2 \\ _2\omega_3 \\ _3v_4 \\ _4\omega_5 \\ _5\omega_6 \end{bmatrix}, \tag{6.5}$$

where $_3v_4$ represents the magnitude of the translational velocity of body 4 relative to body 3.

Now, if the manipulator happens to be in a configuration where the screw coordinates of the six relative twists become linearly dependent, then the matrix J is singular, and its inverse cannot be determined. At this instant, it will not be possible to determine joint velocities that will achieve a desired end effector velocity state $\{^0\omega^6; {}^0v_O^6\}$.[1] Often, when a manipulator approaches a singular configuration, one or more of the joint velocities that are calculated to achieve a desired end effector velocity state will become very large, and the manipulator may all of a sudden move very fast and appear out of control.

Because of this problem, it would be advantageous to identify all the manipulator configurations where the Jacobian matrix J is singular. This way, these configurations can be avoided during path planning. In this chapter, the Jacobian matrix of several six-axis serial industrial manipulators will be evaluated symbolically. The determinant of this matrix will then be expanded, and the conditions that cause the determinant to equal zero will be identified.

6.3   Symbolic Expansion of Screw Coordinates

As presented in Chapter 2, a coordinate system is attached to each link in a standard manner. For link i, the origin of the coordinate system is located at the intersection

---

[1] unless by chance the desired velocity state happens to be a member of the n-space formed by the six screws where n is the rank of J.

of the line along the joint axis $^{i-1}S^i$ and the line along the link vector a^i. The Z axis is parallel to $^{i-1}S^i$ and the X axis is parallel to a^i. The standard coordinate systems attached to links i and j ($j = i + 1$), are shown in Figure 2.25.

Also in Chapter 2, the 4×4 matrix that transforms the coordinates of a point, say point A, from the j to the i coordinate system according to the equation

$$^iP_A = {}^i_jT \; {}^jP_A \tag{6.6}$$

was determined to be

$$^i_jT = \begin{bmatrix} c_{ij} & -s_{ij} & 0 & a_i \\ s_{ij}c_i & c_{ij}c_i & -s_i & -s_i \, _iS_j \\ s_{ij}s_i & c_{ij}s_i & c_i & c_i \, _iS_j \\ 0 & 0 & 0 & 1 \end{bmatrix}, \tag{6.7}$$

where the terms s_{ij} and c_{ij} represent the sine and cosine, respectively, of the joint angle $_i\theta_j$, and s_i and c_i represent the sine and cosine, respectively, of the twist angle α_i. Recall that the columns of the upper left 3×3 submatrix represent the unit direction vectors along the X, Y, and Z axes of the j coordinate system measured with respect to the i coordinate system. The first three elements of the fourth column are the coordinates of the origin of the j coordinate system measured with respect to the i coordinate system.

It is possible to express the origin points and direction vectors of all the link coordinate systems symbolically in terms of a common coordinate system by multiplying sets of 4×4 transformation matrices in an appropriate order. For example, the direction vectors and origin points of the coordinate systems attached to the six moving links of a six-axis serial manipulator may be determined with respect to the fixed (body 0) coordinate system from the third and fourth columns of the following matrices:

$$\begin{aligned} &{}^0_1T, \\ &{}^0_2T = {}^0_1T \, {}^1_2T, \\ &{}^0_3T = {}^0_1T \, {}^1_2T \, {}^2_3T, \\ &{}^0_4T = {}^0_1T \, {}^1_2T \, {}^2_3T \, {}^3_4T, \\ &{}^0_5T = {}^0_1T \, {}^1_2T \, {}^2_3T \, {}^3_4T \, {}^4_5T, \\ &{}^0_6T = {}^0_1T \, {}^1_2T \, {}^2_3T \, {}^3_4T \, {}^4_5T \, {}^5_6T. \end{aligned} \tag{6.8}$$

Recall that the fixed body 0 coordinate system was defined such that its origin was coincident with the origin of the coordinate system attached to the first link. Also, the Z axis of the body 0 coordinate system is parallel to the Z axis of the link 1 coordinate system. Thus, the 4×4 transformation matrix that relates the body 0 and link 1 coordinate systems may be written as

$$^0_1T = \begin{bmatrix} \cos\phi_1 & -\sin\phi_1 & 0 & 0 \\ \sin\phi_1 & \cos\phi_1 & 0 & 0 \\ 0 & 0 & 1 & 0 \\ 0 & 0 & 0 & 1 \end{bmatrix}. \tag{6.9}$$

Symbolically expanding the expressions in (6.8) will result in several large and unwieldy expressions.[2] Further, using these expressions to evaluate the screw coordinates and then the determinant of (6.2) will result in a single expression that if equal to zero will indicate a singular configuration of the manipulator. Attempting to identity the conditions that cause this complicated expression to equal zero will be a huge undertaking for the general case. However, for manipulators with special geometry, such as, for example, three intersecting joint axes or three consecutive parallel joint axes, the symbolic expansion is manageable. Such examples will be shown subsequently in this chapter.

6.4   Selection of Coordinate System

The symbolic determination of singularity conditions can be simplified via a judicious selection of the coordinate system in which to evaluate the screw coordinates. The expressions in (6.8) would be used to express the screw coordinates in the fixed base coordinate system. It is interesting to note that the angle $_5\theta_6$ does not appear in any of the expressions for the directions of the screw axes, since the direction of the last joint axis measured with respect to the fixed coordinate system is independent of this angle.

For this analysis, a coordinate system coincident with the standard coordinate system attached to the third link will be used. The direction vectors and origin points of the coordinate systems attached to the six moving links may be determined with respect to the third (body 3) coordinate system from the third and fourth columns of the following matrices:

$$_1^3T = {_2^3T}\ {_1^2T},$$

$$_2^3T,$$

$$_3^3T = I,$$

$$_4^3T,$$

$$_5^3T = {_4^3T}\ {_5^4T},$$

$$_6^3T = {_4^3T}\ {_5^4T}\ {_6^5T}, \tag{6.10}$$

where the transformation matrix $_i^jT,\ j = i + 1$, is the inverse of $_j^iT$ and is defined as

$$_i^jT = \begin{bmatrix} c_{ij} & s_{ij}c_i & s_{ij}s_i & -c_{ij}a_i \\ -s_{ij} & c_{ij}c_i & c_{ij}s_i & s_{ij}a_i \\ 0 & -s_i & c_i & -{_iS_j} \\ 0 & 0 & 0 & 1 \end{bmatrix}. \tag{6.11}$$

---

[2] Duffy (1980) developed a unified recursive notation to represent commonly occurring patterns of the products of joint and twist angles. With this notation, it is possible to manage the complicated expressions in (6.8). A detailed description of the notation is presented in Duffy (1980) and Crane and Duffy (2008).

As before, the directions of the joint axes and the origin points of the six coordinate systems will be independent of the angle $_5\theta_6$. Further, the first joint angle, $_0\phi_1$, does not appear in any of the expressions in (6.10). Thus, it is apparent that the angles $_0\phi_1$ and $_5\theta_6$ will have no effect as to whether the joint axis screws are linearly dependent and, thus, have no effect as to whether the manipulator is in a singularity configuration.

Symbolically expanding the matrices in (6.10) for the general case in order to obtain symbolic expressions for the directions of the joint axes and for the origin points may still yield unwieldy expressions. However, it will be shown in the next section that for cases of special geometry, as commonly occur for industrial manipulators, the expressions simplify significantly, and it is possible to obtain symbolic expressions for the coordinates of the joint screws and then to symbolically expand the determinant of (6.2) in order to identify all conditions that cause the manipulator to be in a singular configuration.

6.5   Example Problems

6.5.1   Cincinnati Milacron T3-776 Manipulator

A drawing of the kinematic model of the Cincinnati Milacron T3-776 manipulator and the constant mechanism parameters were presented in Problem 3 of Chapter 4. This information is presented again in Figure 6.1 and Table 6.1. In addition, the joint offset $_5S_6$ will be set to 0, which will cause the origin of the coordinate system attached to

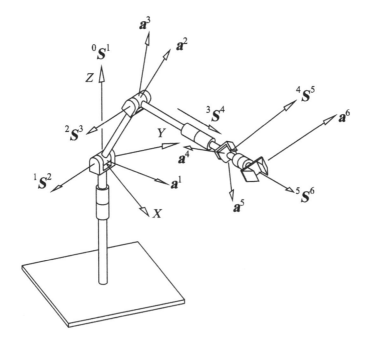

Figure 6.1 Kinematic model of Cincinnati Milacron T3-776 manipulator

Table 6.1. Constant mechanism parameters for T3-776 manipulator

Link Length, in	Twist Angle, deg	Joint Offset, in
$a_1 = 0$	$\alpha_1 = 90$	
$a_2 = 44$	$\alpha_2 = 0$	$_1S_2 = 0$
$a_3 = 0$	$\alpha_3 = 90$	$_2S_3 = 0$
$a_4 = 0$	$\alpha_4 = 61$	$_3S_4 = 55$
$a_5 = 0$	$\alpha_5 = 61$	$_4S_5 = 0$

the last link, link 6, to be coincident with the origin of the coordinate system attached to link 5, since the link length a_5 equals zero.

For this manipulator, the first and second joint axes intersect and are perpendicular, the second and third axes are parallel, the third and fourth axes intersect and are perpendicular, and the last three joint axes intersect. This special geometry greatly simplifies the expressions in (6.10), which are written as

$$_1^3T = \begin{bmatrix} c_{12+23} & 0 & s_{12+23} & -c_{23}a_2 \\ -s_{12+23} & 0 & c_{12+23} & s_{23}a_2 \\ 0 & -1 & 0 & 0 \\ 0 & 0 & 0 & 1 \end{bmatrix}, \tag{6.12}$$

$$_2^3T = \begin{bmatrix} c_{23} & s_{23} & 0 & -c_{23}a_2 \\ -s_{23} & c_{23} & 0 & s_{23}a_2 \\ 0 & 0 & 1 & 0 \\ 0 & 0 & 0 & 1 \end{bmatrix}, \tag{6.13}$$

$$_3^3T = \begin{bmatrix} 1 & 0 & 0 & 0 \\ 0 & 1 & 0 & 0 \\ 0 & 0 & 1 & 0 \\ 0 & 0 & 0 & 1 \end{bmatrix}, \tag{6.14}$$

$$_4^3T = \begin{bmatrix} c_{34} & -s_{34} & 0 & 0 \\ 0 & 0 & -1 & -_3S_4 \\ s_{34} & c_{34} & 0 & 0 \\ 0 & 0 & 0 & 1 \end{bmatrix}, \tag{6.15}$$

$$_5^3T = \begin{bmatrix} K_3 & K_2 & s_{34}s_4 & 0 \\ -s_{45}s_4 & -c_{45}s_4 & -c_4 & -_3S_4 \\ -K_1 & -K_4 & -c_{34}s_4 & 0 \\ 0 & 0 & 0 & 1 \end{bmatrix}, \tag{6.16}$$

$$_6^3T = \begin{bmatrix} A_{11} & A_{12} & -K_2s_5 + s_{34}s_4c_5 & 0 \\ A_{21} & A_{22} & -c_4c_5 + s_4s_5c_{45} & -_3S_4 \\ A_{31} & A_{32} & K_4s_5 - c_{34}s_4c_5 & 0 \\ 0 & 0 & 0 & 1 \end{bmatrix}, \tag{6.17}$$

where the notation s_{12+23} and c_{12+23} is used to represent the sine and cosine, respectively, of the angle $(_1\theta_2 + _2\theta_3)$, and the terms V_{345}, V_{543}, W_{543}, and W_{543}^* are defined as

$$K_1 = -(s_{34}c_{45} + c_{34}s_{45}c_4), \tag{6.18}$$
$$K_2 = -(s_{45}c_{34} + c_{45}s_{34}c_4),$$
$$K_3 = c_{34}c_{45} - s_{34}s_{45}c_4,$$
$$K_4 = s_{34}s_{45} - c_{34}c_{45}c_4.$$

Lastly the terms A_{11}, A_{12}, A_{21}, and A_{22} are expressed as

$$A_{11} = K_3c_{56} + K_2s_{56}c_5 + s_{34}s_{56}s_4s_5, \tag{6.19}$$
$$A_{12} = -K_3s_{56} + K_2c_5c_{56} + s_4s_5s_{34}c_{56},$$
$$A_{21} = -s_4[s_{45}c_{56} + c_{45}s_{56}c_5] - c_4s_5s_{56},$$
$$A_{22} = s_4[s_{45}s_{56} - c_{45}c_{56}c_5] - c_4s_5c_{56},$$
$$A_{31} = -K_1c_{56} - K_4c_5s_{56} - s_4s_5c_{34}s_{56},$$
$$A_{32} = K_1s_{56} - K_4c_5c_{56} - s_4s_5c_{34}c_{56}.$$

Although the terms in (6.19) are complicated, they will not be used when calculating the coordinates of the screw axes along the joints of the manipulator.

The directions of the lines along the six revolute axes, measured with respect to the third coordinate system, can be obtained from the third column of the transformation matrices (6.12) through (6.17) as

$$
{}^0S^1 = \begin{bmatrix} s_{12+23} \\ c_{12+23} \\ 0 \end{bmatrix}, \quad
{}^1S^2 = \begin{bmatrix} 0 \\ 0 \\ 1 \end{bmatrix}, \quad
{}^2S^3 = \begin{bmatrix} 0 \\ 0 \\ 1 \end{bmatrix}, \tag{6.20}
$$

$$
{}^3S^4 = \begin{bmatrix} 0 \\ -1 \\ 0 \end{bmatrix}, \quad
{}^4S^5 = \begin{bmatrix} s_{34}s_4 \\ -c_4 \\ -c_{34}s_4 \end{bmatrix}, \quad
{}^5S^6 = \begin{bmatrix} -K_2s_5 + s_{34}s_4c_5 \\ -c_4c_5 + s_4s_5c_{45} \\ K_4s_5 - c_{34}s_4c_5 \end{bmatrix}.
$$

The coordinates of the origins of each of the six coordinate systems, measured with respect to the third coordinate system, can be obtained from the last column of the transformation matrices (6.12) through (6.17) as

$$
P_1 = \begin{bmatrix} -c_{23}\,a_2 \\ s_{23}\,a_2 \\ 0 \end{bmatrix}, \quad
P_2 = \begin{bmatrix} -c_{23}\,a_2 \\ s_{23}\,a_2 \\ 0 \end{bmatrix}, \quad
P_3 = \begin{bmatrix} 0 \\ 0 \\ 0 \end{bmatrix}, \tag{6.21}
$$

$$
P_4 = \begin{bmatrix} 0 \\ -{}_3S_4 \\ 0 \end{bmatrix}, \quad
P_5 = \begin{bmatrix} 0 \\ -{}_3S_4 \\ 0 \end{bmatrix}, \quad
P_6 = \begin{bmatrix} 0 \\ -{}_3S_4 \\ 0 \end{bmatrix}.
$$

The moments of the lines can be obtained as the cross product of the coordinates of the origin point associated with that line and the direction of the line as

$$ {}^{i-1}S^i_O = P_i \times {}^{i-1}S^i. \tag{6.22} $$

The moment terms are evaluated as

$$
{}^{0}\mathbf{S}_{O}^{1} = \begin{bmatrix} 0 \\ 0 \\ -a_2 c_{12} \end{bmatrix}, \quad {}^{1}\mathbf{S}_{O}^{2} = \begin{bmatrix} a_2 s_{23} \\ a_2 c_{23} \\ 0 \end{bmatrix}, \quad {}^{2}\mathbf{S}_{O}^{3} = \begin{bmatrix} 0 \\ 0 \\ 0 \end{bmatrix}, \tag{6.23}
$$

$$
{}^{3}\mathbf{S}_{O}^{4} = \begin{bmatrix} 0 \\ 0 \\ 0 \end{bmatrix}, \quad {}^{4}\mathbf{S}_{O}^{5} = \begin{bmatrix} {}_{3}S_4 c_{34} s_4 \\ 0 \\ {}_{3}S_4 s_{34} s_4 \end{bmatrix}, \quad {}^{5}\mathbf{S}_{O}^{6} = \begin{bmatrix} -{}_{3}S_4(K_4 s_5 - c_{34} s_4 c_5) \\ 0 \\ {}_{3}S_4(-K_2 s_5 + s_{34} s_4 c_5) \end{bmatrix}.
$$

The Jacobian matrix, whose columns are the coordinates of the lines along the six joint axes may now be written as

$$
J = \begin{bmatrix}
s_{12+23} & 0 & 0 & 0 & s_{34} s_4 & -K_2 s_5 + s_{34} s_4 c_5 \\
c_{12+23} & 0 & 0 & -1 & -c_4 & -c_4 c_5 + s_4 s_5 c_{45} \\
0 & 1 & 1 & 0 & -c_{34} s_4 & K_4 s_5 - c_{34} s_4 c_5 \\
0 & a_2 s_{23} & 0 & 0 & {}_{3}S_4 c_{34} s_4 & -{}_{3}S_4(K_4 s_5 - c_{34} s_4 c_5) \\
0 & a_2 c_{23} & 0 & 0 & 0 & 0 \\
-a_2 c_{12} & 0 & 0 & 0 & {}_{3}S_4 s_{34} s_4 & {}_{3}S_4(-K_2 s_5 + s_{34} s_4 c_5)
\end{bmatrix}. \tag{6.24}
$$

The determinant of this matrix is evaluated as

$$
|J| = {}_{3}S_4\, a_2\, s_4\, s_5\, c_{23}[K_2(\,{}_{3}S_4\, c_{34}\, s_{12+23} + a_2\, c_{12}\, c_{34}) \tag{6.25}
$$
$$
+ K_4(-\,{}_{3}S_4\, s_{34}\, s_{12+23} - a_2\, c_{12}\, s_{34})].
$$

Substituting the definitions of K_2 and K_4 from (6.18) and recognizing that $(s_{34}^2 + c_{34}^2) = 1$ yields

$$
|J| = {}_{3}S_4\, a_2\, s_4\, s_5\, c_{23}\, s_{45}[a_2\, c_{12} + {}_{3}S_4\, s_{12+23}]. \tag{6.26}
$$

It is apparent from this result that the determinant of the Jacobean matrix will equal zero if (i) $\cos\,{}_2\theta_3 = 0$, (ii) $\sin\,{}_4\theta_5 = 0$, or (iii) $a_2\, c_{12} + {}_3 S_4\, s_{34}\, s_{12+23} = 0$. The first case occurs if $ {}_2\theta_3 = \frac{\pi}{2}$ or if $ {}_2\theta_3 = \frac{3\pi}{2}$, which will cause the center of the equivalent ball and socket joint (the point of intersection of the fourth, fifth, and sixth joint axes) to lie in the plane defined by the second and third joint axes. The second case occurs if $ {}_4\theta_5 = 0$, or if $ {}_4\theta_5 = \pi$. When $ {}_4\theta_5 = \pi$, the fourth and sixth joint axes are coincident. When $ {}_4\theta_5 = 0$, the fourth, fifth, and sixth joint axes lie in a common plane. The third case occurs if the center of the equivalent ball and socket joint lies on the line of action of the first joint. These three singularity cases for the T3-776 manipulator are illustrated in Figure 6.2.

6.5.2   G.E. P60 Manipulator

A drawing of the kinematic model of the G.E. P60 manipulator and the constant mechanism parameters are presented in Figure 6.3 and Table 6.2. In addition, the joint offset $ {}_5 S_6$ will be set to 0, which will cause the origin of the coordinate system attached to the last link, link 6, to be coincident with the origin of the coordinate system attached to link 5, since the link length a_5 equals zero. The significant

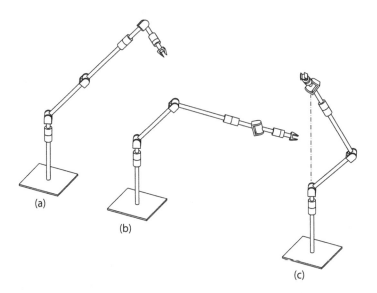

Figure 6.2 Three singularity configurations for the T3-776 manipulator: (a) $_2\theta_3 = \frac{\pi}{2}$, (b) $_4\theta_5 = \pi$, (c) $a_2\,c_{12} + _3S_4\,s_{34}\,s_{12+23} = 0$

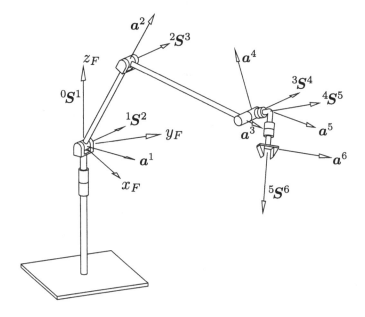

Figure 6.3 Kinematic model of G.E. P60 manipulator

element of the manipulator geometry is that the second, third, and fourth joint axes are parallel.

Substituting the constant mechanism parameters into the expressions in (6.10) yields

Table 6.2. Constant mechanism parameters for G.E. P60 manipulator

Link Length, cm	Twist Angle, deg	Joint Offset, cm
$a_1 = 0$	$\alpha_1 = 270$	
$a_2 = 70$	$\alpha_2 = 0$	$_1S_2 = 0$
$a_3 = 90$	$\alpha_3 = 0$	$_2S_3 = 0$
$a_4 = 0$	$\alpha_4 = 270$	$_3S_4 = 9.8$
$a_5 = 0$	$\alpha_5 = 90$	$_4S_5 = 14.5$

$$
{}_1^3T = \begin{bmatrix} c_{12+23} & 0 & -s_{12+23} & -c_{23}\,a_2 \\ -s_{12+23} & 0 & -c_{12+23} & s_{23}\,a_2 \\ 0 & 1 & 0 & 0 \\ 0 & 0 & 0 & 1 \end{bmatrix}, \tag{6.27}
$$

$$
{}_2^3T = \begin{bmatrix} c_{23} & s_{23} & 0 & -c_{23}\,a_2 \\ -s_{23} & c_{23} & 0 & s_{23}\,a_2 \\ 0 & 0 & 1 & 0 \\ 0 & 0 & 0 & 1 \end{bmatrix}, \tag{6.28}
$$

$$
{}_3^3T = \begin{bmatrix} 1 & 0 & 0 & 0 \\ 0 & 1 & 0 & 0 \\ 0 & 0 & 1 & 0 \\ 0 & 0 & 0 & 1 \end{bmatrix}, \tag{6.29}
$$

$$
{}_4^3T = \begin{bmatrix} c_{34} & -s_{34} & 0 & a_3 \\ s_{34} & c_{34} & 0 & 0 \\ 0 & 0 & 1 & {}_3S_4 \\ 0 & 0 & 0 & 1 \end{bmatrix}, \tag{6.30}
$$

$$
{}_5^3T = \begin{bmatrix} c_{34}c_{45} & -c_{34}s_{45} & -s_{34} & -s_{34}\,{}_4S_5 + a_3 \\ s_{34}c_{45} & -s_{34}s_{45} & c_{34} & c_{34}\,{}_4S_5 \\ -s_{45} & -c_{45} & 0 & {}_3S_4 \\ 0 & 0 & 0 & 1 \end{bmatrix}, \tag{6.31}
$$

$$
{}_6^3T = \begin{bmatrix} c_{34}c_{45}c_{56} - s_{34}s_{56} & -c_{34}c_{45}s_{56} - s_{34}c_{56} & c_{34}s_{45} & -s_{34}\,{}_4S_5 + a_3 \\ s_{34}c_{45}c_{56} + c_{34}s_{56} & -s_{34}c_{45}s_{56} + c_{34}c_{56} & s_{34}s_{45} & c_{34}\,{}_4S_5 \\ -s_{45}c_{56} & s_{45}s_{56} & c_{45} & {}_3S_4 \\ 0 & 0 & 0 & 1 \end{bmatrix}. \tag{6.32}
$$

The directions of the lines along the six revolute axes measured with respect to the third coordinate system can be obtained from the third column of the transformation matrices (6.27) through (6.32) as

$$ {}^{0}\mathbf{S}^{1} = \begin{bmatrix} -s_{12+23} \\ -c_{12+23} \\ 0 \end{bmatrix}, {}^{1}\mathbf{S}^{2} = \begin{bmatrix} 0 \\ 0 \\ 1 \end{bmatrix}, {}^{2}\mathbf{S}^{3} = \begin{bmatrix} 0 \\ 0 \\ 1 \end{bmatrix}, \tag{6.33} $$

$$ {}^{3}\mathbf{S}^{4} = \begin{bmatrix} 0 \\ 0 \\ 1 \end{bmatrix}, {}^{4}\mathbf{S}^{5} = \begin{bmatrix} -s_{34} \\ c_{34} \\ 0 \end{bmatrix}, {}^{5}\mathbf{S}^{6} = \begin{bmatrix} c_{34}s_{45} \\ s_{34}s_{45} \\ c_{45} \end{bmatrix}. $$

The coordinates of the origins of each of the six coordinate systems measured with respect to the third coordinate system can be obtained from the last column of the transformation matrices (6.27) through (6.32) as

$$ \mathbf{P}_{1} = \begin{bmatrix} -c_{23}a_{2} \\ s_{23}a_{2} \\ 0 \end{bmatrix}, \mathbf{P}_{2} = \begin{bmatrix} -c_{23}a_{2} \\ s_{23}a_{2} \\ 0 \end{bmatrix}, \mathbf{P}_{3} = \begin{bmatrix} 0 \\ 0 \\ 0 \end{bmatrix}, \tag{6.34} $$

$$ \mathbf{P}_{4} = \begin{bmatrix} a_{3} \\ 0 \\ {}_{3}S_{4} \end{bmatrix}, \mathbf{P}_{5} = \begin{bmatrix} -{}_{4}S_{5}s_{34} + a_{3} \\ {}_{4}S_{5}c_{34} \\ {}_{3}S_{4} \end{bmatrix}, \mathbf{P}_{6} = \begin{bmatrix} -{}_{4}S_{5}s_{34} + a_{3} \\ {}_{4}S_{5}c_{34} \\ {}_{3}S_{4} \end{bmatrix}. $$

The moments of the lines can be obtained as the cross product of the coordinates of the origin point associated with that line and the direction of the line as

$$ {}^{0}\mathbf{S}_{O}^{1} = \begin{bmatrix} 0 \\ 0 \\ a_{2}c_{12} \end{bmatrix}, {}^{1}\mathbf{S}_{O}^{2} = \begin{bmatrix} a_{2}s_{23} \\ a_{2}c_{23} \\ 0 \end{bmatrix}, {}^{2}\mathbf{S}_{O}^{3} = \begin{bmatrix} 0 \\ 0 \\ 0 \end{bmatrix}, \tag{6.35} $$

$$ {}^{3}\mathbf{S}_{O}^{4} = \begin{bmatrix} 0 \\ -a_{3} \\ 0 \end{bmatrix}, {}^{4}\mathbf{S}_{O}^{5} = \begin{bmatrix} -{}_{3}S_{4}c_{34} \\ -{}_{3}S_{4}s_{34} \\ a_{3}c_{34} \end{bmatrix}, {}^{5}\mathbf{S}_{O}^{6} = \begin{bmatrix} K_{1} \\ K_{2} \\ K_{3} \end{bmatrix}, $$

where

$$ K_{1} = -{}_{3}S_{4}\, s_{34}s_{45} + {}_{4}S_{5}\, c_{34}c_{45}, \tag{6.36} $$

$$ K_{2} = {}_{3}S_{4}\, c_{34}s_{45} + {}_{4}S_{5}\, s_{34}c_{45} - a_{3}c_{45}, $$

$$ K_{3} = a_{3}s_{34}s_{45} - {}_{4}S_{5}\, {}_{4}s_{5}. $$

The Jacobian matrix, whose columns are the coordinates of the lines along the six joint axes, may now be written as

$$ \mathbf{J} = \begin{bmatrix} -s_{12+23} & 0 & 0 & 0 & -s_{34} & c_{34}s_{45} \\ -c_{12+23} & 0 & 0 & 0 & c_{34} & s_{34}s_{45} \\ 0 & 1 & 1 & 1 & 0 & c_{45} \\ 0 & a_{2}s_{23} & 0 & 0 & -{}_{3}S_{4}\,c_{34} & K_{1} \\ 0 & a_{2}c_{23} & 0 & -a_{3} & -{}_{3}S_{4}\,s_{34} & K_{2} \\ a_{2}c_{12} & 0 & 0 & 0 & a_{3}c_{34} & K_{3} \end{bmatrix}. \tag{6.37} $$

The determinant of this matrix is evaluated as

$$ |\mathbf{J}| = a_{2}a_{3}s_{23}s_{45}[a_{2}c_{12} + a_{3}c_{12+23} - {}_{4}S_{5}(s_{12+23+34})], \tag{6.38} $$

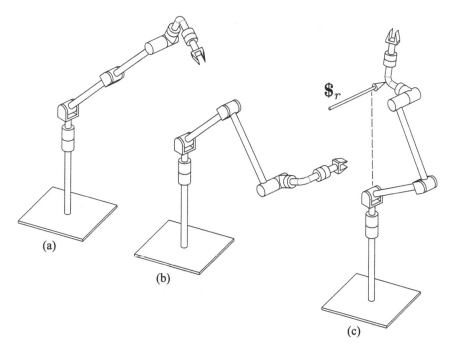

Figure 6.4 Three singular configurations for the G.E. P60 manipulator: (a) $_2\theta_3 = 0$, (b) $_4\theta_5 = 0$, (c) $a_2\,c_{12} + a_3c_{12+23} - {}_4S_5(s_{12+23+34}) = 0$

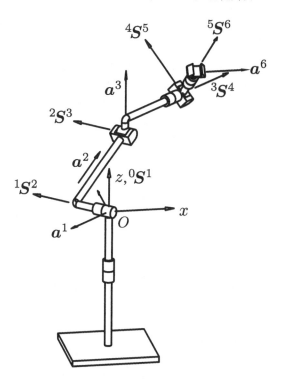

Figure 6.5 Kinematic model of Puma manipulator

Table 6.3. Constant mechanism parameters for PUMA manipulator

Link Length, in	Twist Angle, deg	Joint Offset, in
$a_1 = 0$	$\alpha_1 = 90$	
$a_2 = 17$	$\alpha_2 = 0$	$_1S_2 = 5.9$
$a_3 = 0.8$	$\alpha_3 = 270$	$_2S_3 = 0$
$a_4 = 0$	$\alpha_4 = 90$	$_3S_4 = 17$
$a_5 = 0$	$\alpha_5 = 90$	$_4S_5 = 0$

where $c_{12+23} = \cos(_1\theta_2 + _2\theta_3)$ and $s_{12+23+34} = \sin(_1\theta_2 + _2\theta_3 + _3\theta_4)$.

It is apparent from this result that the determinant of the Jacobean matrix will equal zero if (i) $\sin {_2\theta_3} = 0$, (ii) $\sin {_4\theta_5} = 0$, or (iii) $a_2\, c_{12} + a_3 c_{12+23} - _4S_5(s_{12+23+34}) = 0$. The first case occurs if $_2\theta_3 = 0$, or if $_2\theta_3 = \pi$, which will cause the second, third, and fourth joint axes, which are parallel, to lie in the same plane. The second case occurs if $_4\theta_5 = 0$, or if $_4\theta_5 = \pi$, in which case, the sixth joint axis would become parallel to the second, third, and fourth. When the third condition occurs, there exists a line whose direction is parallel to the direction of the second, third, and fourth joint axes and which intersects the first, fifth, and sixth axes. Thus, a pure force along this line would be simultaneously reciprocal to all six axes of the manipulator. The singularity cases for the G.E. P60 manipulator are illustrated in Figure 6.4.

6.6   Problems

1. The kinematic model of a Puma manipulator is shown in Figure 6.5, and its constant mechanism parameters are listed in Table 6.3. Identify all singular configurations of this manipulator.

7 Acceleration Analysis of Serially Connected Rigid Bodies

> Acceleration is much more subtle and complex in
> its nature than the velocity change from which it comes.
>
> Hartenberg and Denavit (1964)

7.1 Introduction

The objectives of this chapter are twofold. First, it will be necessary to identify parameters from which the acceleration of any point in a body can be determined relative to a reference body. These parameters will be called the acceleration state parameters. The second objective relates to the analysis of a series of rigid bodies that are interconnected by a sequence of simple joints, i.e., revolute, prismatic, or screw pairs. Expressions for the acceleration state parameters of the last link of the chain will be developed in terms of the angular acceleration (or linear acceleration for a prismatic joint) of link $i+1$ relative to link i. From these equations, it will be possible to perform forward and reverse acceleration analyses for serial manipulators.

7.2 General Case of Two Bodies

Figure 7.1 shows two bodies and a reference coordinate system that is embedded in body 0, which is fixed. At this instant, the velocity state of body 1 relative to body 0 is assumed known, i.e., $^0\omega^1$ and $^0v_O^1$. Also, at this instant, the angular acceleration of body 1 relative to body 0, $^0\alpha^1$, and the linear acceleration of the point in body 1 that is coincident with the origin of the reference frame, $^0a_O^1$, are assumed to be known. The objective is to determine the linear acceleration of an arbitrary point P in body 1 in terms of the given parameters. If this can be accomplished, then the terms $^0\alpha^1$ and $^0a_O^1$ will be defined as the acceleration state parameters of body 1 relative to body 0.

The velocity of point P in body 1 was expressed in terms of the velocity state parameters in equation (4.41) as

$$^0v_P^1 = {}^0v_O^1 + {}^0\omega^1 \times {}^0r_{O1 \to P1}. \tag{7.1}$$

Here, the notation $r_{O1 \to P1}$ is used in that it is important to realize, before taking derivatives, that the vector being referred to originates at point O in body 1 and

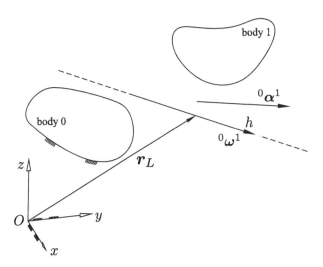

Figure 7.1 Body 1 moving with respect to body 0

terminates at point P, also in body 1. Taking a derivative of this equation with respect to body 0 yields

$$^0\frac{d}{dt}{}^0v_P^1 = {}^0\frac{d}{dt}{}^0v_O^1 + {}^0\frac{d}{dt}\left({}^0\omega^1 \times {}^0r_{O1 \to P1}\right). \tag{7.2}$$

The derivative of the velocity of a point with respect to body 0, where the velocity is measured in terms of body 0, is simply the acceleration of that point with respect to body 0. Thus, (7.2) may be written as

$$^0a_P^1 = {}^0a_O^1 + {}^0\frac{d}{dt}\left({}^0\omega^1 \times {}^0r_{O1 \to P1}\right). \tag{7.3}$$

The chain rule may be used to evaluate the last term as

$$^0a_P^1 = {}^0a_O^1 + \left({}^0\frac{d}{dt}{}^0\omega^1\right) \times {}^0r_{O1 \to P1} + {}^0\omega^1 \times \left({}^0\frac{d}{dt}{}^0r_{O1 \to P1}\right). \tag{7.4}$$

The angular acceleration vector is defined as the derivative of the angular velocity vector and, thus, ${}^0\frac{d}{dt}{}^0\omega^1 = {}^0\alpha^1$. Also, as shown in Section 4.2 (Case 3) the derivative of a vector whose head and tail both lie in body 1 may be written as

$$^0\frac{d}{dt}{}^0r_{O1 \to P1} = {}^0\omega^1 \times {}^0r_{O \to P}, \tag{7.5}$$

where now the subscript 1 has been deleted since point O in body 1 is coincident with that in body 0. Now, (7.4) may be written as

$$^0a_P^1 = {}^0a_O^1 + {}^0\alpha^1 \times {}^0r_{O \to P} + {}^0\omega^1 \times \left({}^0\omega^1 \times {}^0r_{O \to P}\right). \tag{7.6}$$

It is shown by (7.6) that the acceleration of any point in the moving body can be calculated based on knowledge of the velocity state, $\{{}^0\omega^1, {}^0v_O^1\}$, the parameters ${}^0\alpha^1$ and ${}^0a_O^1$, and the coordinates of point P, i.e., ${}^0r_{O \to P}$. Thus, the parameters ${}^0\alpha^1$ and

$^0a^1_O$ meet the requirement to be referred to as the acceleration state parameters of body 1 relative to body 0.

7.3   General Case of Three Bodies

Figure 7.2 shows three bodies. Body 0 may be considered as being fixed, while bodies 1 and 2 are in motion. The objective is to determine the acceleration state parameters of body 2 relative to body 0, assuming that the velocity and acceleration state parameters of body 2 are known relative to body 1, as well as the velocity and acceleration parameters of body 1 relative to body 0. It is important to note that all the velocity and acceleration state parameters are expressed in terms of coordinate systems that at this instant are coincident and aligned. Specifically, the terms $\{^0\omega^1; {}^0v^1_O\}$, $\{^0\alpha^1; {}^0a^1_O\}$, $\{^1\omega^2; {}^1v^2_O\}$, and $\{^1\alpha^2; {}^1a^2_O\}$ are given, and the objective is to determine the acceleration state parameters $^0\alpha^2$ and $^0a^2_O$.

The angular acceleration of body 2 relative to body 0 may be written as

$$^0\alpha^2 = {}^0\frac{d}{dt}{}^0\omega^2. \tag{7.7}$$

From (4.65), $^0\omega^2 = {}^0\omega^1 + {}^1\omega^2$, and (7.7) may now be written as

$$^0\alpha^2 = {}^0\frac{d}{dt}\left(^0\omega^1 + {}^1\omega^2\right), \tag{7.8}$$

which may be expanded as

$$^0\alpha^2 = {}^0\frac{d}{dt}{}^0\omega^1 + {}^0\frac{d}{dt}{}^1\omega^2. \tag{7.9}$$

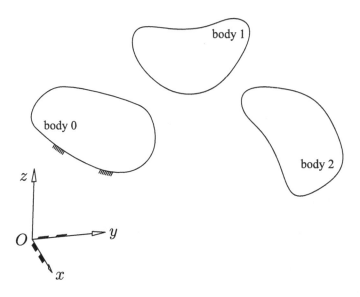

Figure 7.2 Three bodies

The first term is simply the angular acceleration of body 1 relative to body 0 and, thus,

$$^0\alpha^2 = {}^0\alpha^1 + {}^0\frac{d}{dt}{}^1\omega^2. \tag{7.10}$$

Equation (4.34) can be used to express the derivative of $^1\omega^2$ with respect to body 0 in terms of its derivative with respect to body 1 as

$$^0\frac{d}{dt}{}^1\omega^2 = {}^1\frac{d}{dt}{}^1\omega^2 + {}^0\omega^1 \times {}^1\omega^2. \tag{7.11}$$

The first term on the right side of this equation is the derivative of the angular velocity of body 2 relative to body 1 taken with respect to body 1. This will clearly equal the angular acceleration of body 2 relative to body 1, and (7.10) may now be written as

$$^0\alpha^2 = {}^0\alpha^1 + {}^1\alpha^2 + {}^0\omega^1 \times {}^1\omega^2. \tag{7.12}$$

Thus, the angular acceleration of body 2 with respect to body 0 is not simply the sum of the the angular acceleration of body 1 relative to body 0 and the angular acceleration of body 2 relative to body 1.

It remains to determine the acceleration of the point in body 2 that is coincident with the origin of the reference system, i.e., $^0a_O^2$. Figure 7.3 shows the three bodies together with points O_0, B_1, and P_2. The subscripts associated with the point names indicate that we are referring to point O in body 0, point B in body 1, and point P in body 2. The vector from O_0 to P_2 may be expressed by the following summation:

$$r_{O0 \to P2} = r_{O0 \to B1} + r_{B1 \to P2}. \tag{7.13}$$

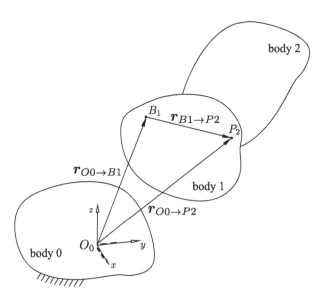

Figure 7.3 Three bodies

Taking a derivative with respect to body 0 yields

$$\sideset{^0}{}{\frac{d}{dt}} r_{O0 \to P2} = \sideset{^0}{}{\frac{d}{dt}} r_{O0 \to B1} + \sideset{^0}{}{\frac{d}{dt}} r_{B1 \to P2}. \tag{7.14}$$

Since the first two vectors originate at a point in body 0, and the derivative is taken with respect to body 0, the first two terms may be interpreted as the velocity of the end point of the vector with respect to body 0. Thus,

$$\sideset{^0}{}{v_P^2} = \sideset{^0}{}{v_B^1} + \sideset{^0}{}{\frac{d}{dt}} r_{B1 \to P2}. \tag{7.15}$$

Equation (4.26) can be used to relate the derivative of the vector from point B_1 to point P_2 taken with respect to body 0 to the derivative taken with respect to body 1 as

$$\sideset{^0}{}{\frac{d}{dt}} r_{B1 \to P2} = \sideset{^1}{}{\frac{d}{dt}} r_{B1 \to P2} + \sideset{^0}{}{\omega^1} \times r_{B1 \to P2}. \tag{7.16}$$

Recognizing that $\sideset{^1}{}{\frac{d}{dt}} r_{B1 \to P2} = \sideset{^1}{}{v_P^2}$ and substituting (7.16) into (7.15) yields

$$\sideset{^0}{}{v_P^2} = \sideset{^0}{}{v_B^1} + \sideset{^1}{}{v_P^2} + \sideset{^0}{}{\omega^1} \times r_{B1 \to P2}. \tag{7.17}$$

Applying another derivative with respect to body 0 gives

$$\sideset{^0}{}{\frac{d}{dt}} \sideset{^0}{}{v_P^2} = \sideset{^0}{}{\frac{d}{dt}} \sideset{^0}{}{v_B^1} + \sideset{^0}{}{\frac{d}{dt}} \sideset{^1}{}{v_P^2} + \sideset{^0}{}{\frac{d}{dt}} \left(\sideset{^0}{}{\omega^1} \times r_{B1 \to P2} \right). \tag{7.18}$$

The first two terms represent the acceleration of point P and point B with respect to body 0 and, thus,

$$\sideset{^0}{}{a_P^2} = \sideset{^0}{}{a_B^1} + \sideset{^0}{}{\frac{d}{dt}} \sideset{^1}{}{v_P^2} + \sideset{^0}{}{\frac{d}{dt}} \left(\sideset{^0}{}{\omega^1} \times r_{B1 \to P2} \right). \tag{7.19}$$

Again, (4.34) can be used to relate the derivative of the velocity of point P_2 taken with respect to body 0 to the derivative taken with respect to body 1 as

$$\sideset{^0}{}{\frac{d}{dt}} \sideset{^1}{}{v_P^2} = \sideset{^1}{}{\frac{d}{dt}} \sideset{^1}{}{v_P^2} + \sideset{^0}{}{\omega^1} \times \sideset{^1}{}{v_P^2}, \tag{7.20}$$

which may be written as

$$\sideset{^0}{}{\frac{d}{dt}} \sideset{^1}{}{v_P^2} = \sideset{^1}{}{a_P^2} + \sideset{^0}{}{\omega^1} \times \sideset{^1}{}{v_P^2}. \tag{7.21}$$

Substituting (7.21) into (7.19) gives

$$\sideset{^0}{}{a_P^2} = \sideset{^0}{}{a_B^1} + \sideset{^1}{}{a_P^2} + \sideset{^0}{}{\omega^1} \times \sideset{^1}{}{v_P^2} + \sideset{^0}{}{\frac{d}{dt}} \left(\sideset{^0}{}{\omega^1} \times r_{B1 \to P2} \right). \tag{7.22}$$

The last term in this equation may be expanded as

$$\sideset{^0}{}{\frac{d}{dt}} \left(\sideset{^0}{}{\omega^1} \times r_{B1 \to P2} \right) = \left(\sideset{^0}{}{\frac{d}{dt}} \sideset{^0}{}{\omega^1} \right) \times r_{B1 \to P2} \tag{7.23}$$

$$+ \sideset{^0}{}{\omega^1} \times \left(\sideset{^0}{}{\frac{d}{dt}} r_{B1 \to P2} \right).$$

Substituting $^0\frac{d}{dt}\,^0\omega^1 = {}^0\alpha^1$ and using (4.26) to evaluate the derivative of the vector from point B_1 to point P_2 in terms of body 1 gives

$$^0\frac{d}{dt}\left(^0\omega^1 \times r_{B1\rightarrow P2}\right) = {}^0\alpha^1 \times r_{B1\rightarrow P2} \tag{7.24}$$

$$+ {}^0\omega^1 \times \left(^1\frac{d}{dt}r_{B1\rightarrow P2} + {}^0\omega^1 \times r_{B1\rightarrow P2}\right).$$

Again, recognizing that $^1\frac{d}{dt}r_{B1\rightarrow P2} = {}^1v_P^2$ and substituting (7.24) into (7.22) yields

$$^0a_P^2 = {}^0a_B^1 + {}^1a_P^2 + {}^0\omega^1 \times {}^1v_P^2 + {}^0\alpha^1 \times r_{B1\rightarrow P2} \tag{7.25}$$

$$+ {}^0\omega^1 \times \left(^1v_P^2 + {}^0\omega^1 \times r_{B1\rightarrow P2}\right).$$

This result may be rewritten as

$$^0a_P^2 = {}^0a_B^1 + {}^1a_P^2 + 2\,^0\omega^1 \times {}^1v_P^2 + {}^0\alpha^1 \times r_{B1\rightarrow P2} + {}^0\omega^1 \times \left(^0\omega^1 \times r_{B1\rightarrow P2}\right). \tag{7.26}$$

Now, consider the case where point B_1 is coincident with point P_2, i.e., $r_{B1\rightarrow P2} = 0$. Equation (7.26) may now be written as

$$^0a_P^2 = {}^0a_P^1 + {}^1a_P^2 + 2\,^0\omega^1 \times {}^1v_P^2, \tag{7.27}$$

and this equation relates the acceleration of point P in body 2 (as seen with respect to body 0) to the acceleration of point P embedded in body 1 (as seen with respect to body 0) and the acceleration of point P embedded in body 2 (as seen with respect to body 1). Point P is completely arbitrary, and the expression in (7.27) can apply to any point that is coincident in bodies 1 and 2. Applying the result to the points in bodies 1 and 2 that are coincident with the origin of the reference frame gives

$$^0a_O^2 = {}^0a_O^1 + {}^1a_O^2 + 2\,^0\omega^1 \times {}^1v_O^2. \tag{7.28}$$

Equations (7.12) and (7.28) define the acceleration state parameters of body 2 relative to body 0 in terms of the velocity and acceleration states of body 2 relative to body 1 and body 1 relative to body 0.

7.4   General Case of Four Bodies

Figure 7.4 shows four bodies. Body 0 may be considered as being fixed, while bodies 1, 2, and 3 are in motion. The velocity and acceleration states of body $i+1$ are assumed known with respect to body i, $i = 0\ldots2$. Again, it is important to note that all the velocity and acceleration state parameters are expressed in terms of coordinate systems that at this instant are coincident and aligned. The objective here is to determine the acceleration state parameters of body 3 relative to body 0, i.e., $^0\alpha^3$ and $^0a_O^3$.

The angular acceleration of body 3 with respect to body 0 is defined as

$$^0\alpha^3 = {}^0\frac{d}{dt}\left(^0\omega^3\right). \tag{7.29}$$

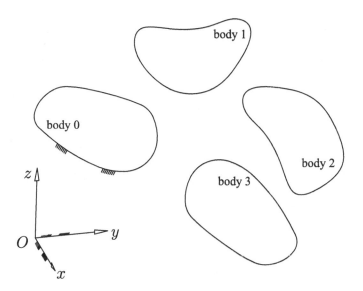

Figure 7.4 Four bodies

From (4.79), this may be written as

$$^0\alpha^3 = {}^0\frac{d}{dt}\left({}^0\omega^1 + {}^1\omega^2 + {}^2\omega^3\right), \tag{7.30}$$

which may be expanded as

$$^0\alpha^3 = {}^0\frac{d}{dt}\,{}^0\omega^1 + {}^0\frac{d}{dt}\,{}^1\omega^2 + {}^0\frac{d}{dt}\,{}^2\omega^3. \tag{7.31}$$

The first term on the right side of this equation is obviously $^0\alpha^1$. The remaining two derivatives will be re-expressed in terms of their derivative with respect to body 1 by applying (4.34) to give

$$^0\alpha^3 = {}^0\alpha^1 + \left({}^1\frac{d}{dt}\,{}^1\omega^2 + {}^0\omega^1 \times {}^1\omega^2\right) + \left({}^1\frac{d}{dt}\,{}^2\omega^3 + {}^0\omega^1 \times {}^2\omega^3\right). \tag{7.32}$$

Recognizing that $^1\frac{d}{dt}\,{}^1\omega^2 = {}^1\alpha^2$ and rearranging gives

$$^0\alpha^3 = {}^0\alpha^1 + {}^1\alpha^2 + {}^0\omega^1 \times \left({}^1\omega^2 + {}^2\omega^3\right) + {}^1\frac{d}{dt}\,{}^2\omega^3. \tag{7.33}$$

Now, using (4.34) to express the remaining derivative in terms of a derivative with respect to body 2 yields

$$^0\alpha^3 = {}^0\alpha^1 + {}^1\alpha^2 + {}^0\omega^1 \times \left({}^1\omega^2 + {}^2\omega^3\right) + \left({}^2\frac{d}{dt}\,{}^2\omega^3 + {}^1\omega^2 \times {}^2\omega^3\right). \tag{7.34}$$

Lastly, recognizing that $^2\frac{d}{dt}\,{}^2\omega^3 = {}^2\alpha^3$ and rearranging gives

$$^0\alpha^3 = {}^0\alpha^1 + {}^1\alpha^2 + {}^2\alpha^3 + {}^0\omega^1 \times \left({}^1\omega^2 + {}^2\omega^3\right) + {}^1\omega^2 \times {}^2\omega^3. \tag{7.35}$$

It remains to determine the acceleration state parameter $^{0}a_{O}^{3}$. Equation (7.28) can be used to determine the acceleration of the point in body 3 that is coincident with the origin of the reference frame with respect to body 1 by adding one to each of the superscripts to give

$$^{1}a_{O}^{3} = {}^{1}a_{O}^{2} + {}^{2}a_{O}^{3} + 2\,{}^{1}\omega^{2} \times {}^{2}v_{O}^{3}. \tag{7.36}$$

Now, consider that there are only three bodies, i.e., bodies 0, 1, and 3. The acceleration of the point in the distal body, body 3 in this case, that is coincident with the origin, measured with respect to the first of our "three bodies," body 0 in this case, was determined in (7.28) to be

$$^{0}a_{O}^{3} = {}^{0}a_{O}^{1} + {}^{1}a_{O}^{3} + 2\,{}^{0}\omega^{1} \times {}^{1}v_{O}^{3}. \tag{7.37}$$

Substituting (7.36) into (7.37) and substituting $^{1}v_{O}^{3} = {}^{1}v_{O}^{2} + {}^{2}v_{O}^{3}$ gives the result

$$^{0}a_{O}^{3} = {}^{0}a_{O}^{1} + {}^{1}a_{O}^{2} + {}^{2}a_{O}^{3} + 2\,{}^{0}\omega^{1} \times \left({}^{1}v_{O}^{2} + {}^{2}v_{O}^{3}\right) + 2\,{}^{1}\omega^{2} \times {}^{2}v_{O}^{3}. \tag{7.38}$$

7.5   General Case of *n* Bodies

The approach undertaken in the previous section can be extended for the case of $n+1$ bodies, numbered from 0 to n, interconnected by n screw joints. It can readily be proved by induction that for this case

$$^{0}\alpha^{n} = {}^{0}\alpha^{1} + {}^{1}\alpha^{2} + {}^{2}\alpha^{3} + \cdots + {}^{n-1}\alpha^{n} \tag{7.39}$$
$$+ {}^{0}\omega^{1} \times \left({}^{1}\omega^{2} + {}^{2}\omega^{3} + \cdots + {}^{n-1}\omega^{n}\right)$$
$$+ {}^{1}\omega^{2} \times \left({}^{2}\omega^{3} + \cdots + {}^{n-1}\omega^{n}\right)$$
$$\vdots$$
$$+ {}^{n-2}\omega^{n-1} \times \left({}^{n-1}\omega^{n}\right)$$

and

$$^{0}a_{O}^{n} = {}^{0}a_{O}^{1} + {}^{1}a_{O}^{2} + {}^{2}a_{O}^{3} + \cdots + {}^{n-1}a_{O}^{n} \tag{7.40}$$
$$+ 2\,{}^{0}\omega^{1} \times \left({}^{1}v_{O}^{2} + {}^{2}v_{O}^{3} + \cdots + {}^{n-1}v_{O}^{n}\right)$$
$$+ 2\,{}^{1}\omega^{2} \times \left({}^{2}v_{O}^{3} + \cdots + {}^{n-1}v_{O}^{n}\right)$$
$$\vdots$$
$$+ 2\,{}^{n-2}\omega^{n-1} \times \left({}^{n-1}v_{O}^{n}\right).$$

For the case of seven bodies, numbered 0 to 6, interconnected by six screw joints (which is the case for a typical six axis robot manipulator), these equations may be written as

$$^0\alpha^6 = {}^0\alpha^1 + {}^1\alpha^2 + {}^2\alpha^3 + {}^3\alpha^4 + {}^4\alpha^5 + {}^5\alpha^6 \tag{7.41}$$
$$+ {}^0\omega^1 \times \left({}^1\omega^2 + {}^2\omega^3 + {}^3\omega^4 + {}^4\omega^5 + {}^5\omega^6\right)$$
$$+ {}^1\omega^2 \times \left({}^2\omega^3 + {}^3\omega^4 + {}^4\omega^5 + {}^5\omega^6\right)$$
$$+ {}^2\omega^3 \times \left({}^3\omega^4 + {}^4\omega^5 + {}^5\omega^6\right)$$
$$+ {}^3\omega^4 \times \left({}^4\omega^5 + {}^5\omega^6\right)$$
$$+ {}^4\omega^5 \times \left({}^5\omega^6\right)$$

and

$$^0a_O^6 = {}^0a_O^1 + {}^1a_O^2 + {}^2a_O^3 + {}^3a_O^4 + {}^4a_O^5 + {}^5a_O^6 \tag{7.42}$$
$$+ 2\,{}^0\omega^1 \times \left({}^1v_O^2 + {}^2v_O^3 + {}^3v_O^4 + {}^4v_O^5 + {}^5v_O^6\right)$$
$$+ 2\,{}^1\omega^2 \times \left({}^2v_O^3 + {}^3v_O^4 + {}^4v_O^5 + {}^5v_O^6\right)$$
$$+ 2\,{}^2\omega^3 \times \left({}^3v_O^4 + {}^4v_O^5 + {}^5v_O^6\right)$$
$$+ 2\,{}^3\omega^4 \times \left({}^4v_O^5 + {}^5v_O^6\right)$$
$$+ 2\,{}^4\omega^5 \times \left({}^5v_O^6\right).$$

7.6 Two Bodies Connected by a Screw Joint

Figure 7.5 shows body 1, which is moving about a screw joint relative to body 0. What is different, compared to the previous cases, is that the direction of the angular velocity vector will not change with respect to the reference system. Thus, the direction of the angular acceleration vector must be parallel to the direction of the angular velocity vector. In this way, the angular acceleration acts to change the magnitude of the angular velocity, but not its direction.

For this case, the following information can be assumed to be known:

- $^0S^1$, the direction along the line of action of the screw joint,
- $_0\omega_1$, the magnitude of the angular velocity of body 1 relative to body 0,
- $_0\alpha_1$, the magnitude of the angular acceleration of body 1 relative to body 0,
- $^0S_{OL}^1$, the moment of the line of action of the screw joint relative to the reference frame, and
- h, the pitch of the screw joint.

The objective is to determine the acceleration state parameters of body 1 relative to body 0, i.e., $^0\alpha^1$ and $^0a_O^1$. The solution for $^0\alpha^1$ is trivial. It is apparent that

$$^0\alpha^1 = {}_0\alpha_1\,{}^0S^1. \tag{7.43}$$

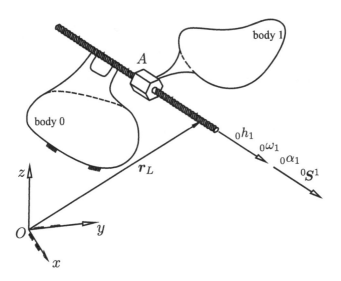

Figure 7.5 Body 1 connected to body 0 by a screw joint

The solution for $^0a_O^1$ will begin by determining the acceleration of point A, which is embedded in body 1 and lies on the line of action of the screw joint. The velocity of point A may be written as

$$^0v_A^1 = {}^0v_O^1 + {}^0\omega^1 \times p_A.\tag{7.44}$$

The velocity of the point in body 1 that is coincident with the origin may be written as

$$^0v_O^1 = {}_0\omega_1 \left[(r \times {}^0S^1) + h\,{}^0S^1 \right],\tag{7.45}$$

where r is any arbitrary point on the line of action of the screw. Substituting (7.45) into (7.44) and recognizing that ${}_0\omega_1\,{}^0S^1 = {}^0\omega^1$ gives

$$^0v_A^1 = (r \times {}^0\omega^1) + {}_0\omega_1\,h\,{}^0S^1 + {}^0\omega^1 \times p_A.$$

This may be regrouped as

$$^0v_A^1 = {}^0\omega^1 \times (p_A - r) + {}_0\omega_1\,h\,{}^0S^1.\tag{7.46}$$

The vector $(p_A - r)$ is parallel to ${}^0\omega^1$, and the velocity of point A is due solely to the pitch of the screw and may be written as

$$^0v_A^1 = {}_0\omega_1\,{}_0h_1\,{}^0S^1.\tag{7.47}$$

The acceleration of point A is obtained by taking a time derivative of this equation with respect to body 0. Since the pitch and direction of the line of action of the screw are invariant, the acceleration of point A can be written as

$$^0a_A^1 = {}_0\alpha_1\,{}_0h_1\,{}^0S^1.\tag{7.48}$$

The acceleration of an arbitrary point in body 1 was written in (7.6) in terms of the acceleration of the point in body 1 that is coincident with the origin. Using this result, the acceleration of point A may be written as

$$^0 a_A^1 = {}^0 a_O^1 + {}^0 \alpha^1 \times {}^0 r_{O \to A} + {}^0 \omega^1 \times \left({}^0 \omega^1 \times {}^0 r_{O \to A} \right). \tag{7.49}$$

Substituting (7.43) and (7.48) into (7.49) and solving for $^0 a_O^1$ gives

$$^0 a_O^1 = {}_0\alpha_1 {}_0 h_1 {}^0 S^1 - {}_0\alpha_1 {}^0 S^1 \times {}^0 r_{O \to A} - {}^0 \omega^1 \times \left({}^0 \omega^1 \times {}^0 r_{O \to A} \right). \tag{7.50}$$

This equation can be regrouped as

$$^0 a_O^1 = {}_0\alpha_1 \left({}^0 r_{O \to A} \times {}^0 S^1 + {}_0 h_1 {}^0 S^1 \right) - {}^0 \omega^1 \times \left({}^0 \omega^1 \times {}^0 r_{O \to A} \right). \tag{7.51}$$

Recognizing that $^0 r_{O \to A} \times {}^0 S^1 + {}_0 h_1 {}^0 S^1 = {}^0 S_O^1$ gives

$$^0 a_O^1 = {}_0\alpha_1 {}^0 S_O^1 - {}^0 \omega^1 \times \left({}^0 \omega^1 \times {}^0 r_{O \to A} \right). \tag{7.52}$$

This result may now be rearranged as

$$^0 a_O^1 = {}_0\alpha_1 {}^0 S_O^1 + {}^0 \omega^1 \times \left({}^0 r_{O \to A} \times {}^0 \omega^1 \right). \tag{7.53}$$

Recognizing that $^0 r_{O \to A} \times {}^0 \omega^1 = {}_0 \omega_1 {}^0 S_{OL}^1$ gives

$$^0 a_O^1 = {}_0\alpha_1 {}^0 S_O^1 + {}^0 \omega^1 \times {}_0 \omega_1 {}^0 S_{OL}^1. \tag{7.54}$$

Since $^0 S_{OL}^1 = {}^0 S_O^1 - h {}^0 S^1$, and since $^0 \omega^1 \parallel {}^0 S^1$, it is apparent that $^0 \omega^1 \times {}^0 S_{OL}^1 = {}^0 \omega^1 \times {}^0 S_O^1$. Substituting this result into (7.54) yields

$$^0 a_O^1 = {}_0\alpha_1 {}^0 S_O^1 + {}^0 \omega^1 \times {}_0 \omega_1 {}^0 S_O^1. \tag{7.55}$$

Lastly, recognizing that $_0\omega_1 {}^0 S_O^1 = {}^0 v_O^1$ gives

$$^0 a_O^1 = {}_0\alpha_1 {}^0 S_O^1 + {}^0 \omega^1 \times {}^0 v_O^1. \tag{7.56}$$

Now that the acceleration state parameters $^0 \alpha^1$ and $^0 a_O^1$ have been identified, (7.6) may be used to determine the acceleration of any point in body 1.

7.7 Two Bodies Connected by a Revolute Joint

Figure 7.6 shows body 1, which is moving about a revolute joint relative to body 0. As in the previous section, the direction of the angular velocity vector will not change with respect to the reference system and, thus, the direction of the angular acceleration vector must be parallel to the direction of the angular velocity vector.

For this case, the following information can be assumed to be known:

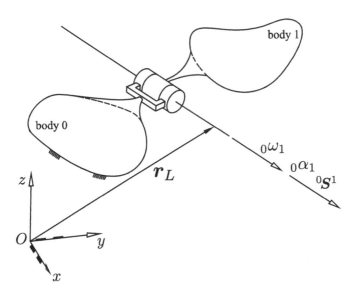

Figure 7.6 Body 1 connected to body 0 by a revolute joint

- $^0S^1$, the direction along the line of action of the screw joint,
- $_0\omega_1$, the magnitude of the angular velocity of body 1 relative to body 0,
- $_0\alpha_1$, the magnitude of the angular acceleration of body 1 relative to body 0, and
- $^0S^1_{OL}$, the moment of the line of action of the revolute joint relative to the reference frame.

The objective is to determine the acceleration state parameters of body 1 relative to body 0, i.e., $^0\alpha^1$ and $^0a^1_O$. The solution will be the same as in the previous section, with the exception that now the pitch equals zero. Thus,

$$^0\alpha^1 = {}_0\alpha_1\,{}^0S^1 \tag{7.57}$$

and

$$^0a^1_O = {}_0\alpha_1\,{}^0S^1_{OL} + {}^0\omega^1 \times {}^0v^1_O, \tag{7.58}$$

where for the revolute joint $^0\omega^1 \cdot {}^0v^1_O = 0$.

7.8   Two Bodies Connected by a Prismatic Joint

Figure 7.7 shows body 1, which is moving in translation relative to body 0. For this case, the direction of the translation, $^0S^1$, is assumed known, together with the magnitude of the translational velocity, $_0v_1$, and the translational acceleration, $_0a_1$. As in the previous cases, the objective is to determine the acceleration state parameters of body 1 relative to body 0.

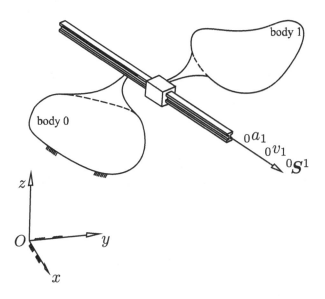

Figure 7.7 Body 1 connected to body 0 by a prismatic joint

    The solution to this problem is trivial in that all points in body 1 will have the same velocity and the same acceleration with respect to body 0. Thus, for this case

$$^0\alpha^1 = 0, \tag{7.59}$$

$$^0a^1_O = {}_0a_1\,{}^0S^1. \tag{7.60}$$

7.9   Forward Acceleration Analysis for Serial Manipulators

    In the forward acceleration analysis of a six-axis serial manipulator, the individual joint angles, $_k\theta_{k+1}$, (or joint offsets for a prismatic joint, $_kS_{k-1}$), the individual joint angular velocities, $_k\omega_{k+1}$ (or linear velocity for a prismatic joint, $_kv_{k+1}$), and the individual joint angular accelerations, $_k\alpha_{k+1}$ (or linear accelerations for a prismatic joint, $_ka_{k+1}$), $k = 0\dots5$, are assumed known, in addition to the constant mechanism parameters. It is desired to obtain the acceleration state parameters of the last link of the robot relative to ground, i.e., $^0\alpha^6$ and $^0a^6_O$.

    Since all the joint angles are known, the Plücker line coordinates of all the joint axes of the manipulator can be determined, i.e., $\{{}^kS^{k+1}; {}^kS^{k+1}_{OL}\}$, $k = 0\dots5$. Further, the coordinates of each of the six screws, $\{{}^kS^{k+1}; {}^kS^{k+1}_O\}$, $k = 0\dots5$, can be determined by by incorporating the pitch term of each axis. Since the joint angular velocities are known, the instantaneous twist of body k relative to body $k + 1$ may be obtained as $\{{}^k\omega^{k+1}; {}^kv^{k+1}_O\} = {}_k\omega_{k+1}\{{}^kS^{k+1}; {}^kS^{k+1}_O\}$, $k = 0\dots5$. Lastly, $^k\alpha^{k+1}$ and $^ka^{k+1}_O$ may be obtained from (7.43) and (7.56) (for a screw joint) or from (7.59) and (7.60) for a prismatic joint. Equations (7.41) and (7.42) can now be used to directly solve for $^0\alpha^6$ and $^0a^6_O$, as all the terms on the right side of these equations are known.

7.9.1   Sample Problem

Section 4.7.1 presented a numerical example of a reverse velocity analysis for a Puma industrial robot. For the particular configuration under consideration, the Plücker coordinates of the lines along each of the six revolute joints of the manipulator were found to be

$$
\begin{bmatrix} {}^0 S^1 \\ {}^0 S_O^1 \end{bmatrix} = \begin{bmatrix} 0 \\ 0 \\ 1 \\ 0 \\ 0 \\ 0 \end{bmatrix}, \quad
\begin{bmatrix} {}^1 S^2 \\ {}^1 S_O^2 \end{bmatrix} = \begin{bmatrix} -0.7071 \\ 0.7071 \\ 0 \\ 0 \\ 0 \\ 0 \end{bmatrix}, \quad
\begin{bmatrix} {}^2 S^3 \\ {}^2 S_O^3 \end{bmatrix} = \begin{bmatrix} -0.7071 \\ 0.7071 \\ 0 \\ -6.0104 \\ -6.0104 \\ 14.7224 \end{bmatrix}, \quad (7.61)
$$

$$
\begin{bmatrix} {}^3 S^4 \\ {}^3 S_O^4 \end{bmatrix} = \begin{bmatrix} 0.7071 \\ 0.7071 \\ 0 \\ -6.5761 \\ 6.5761 \\ -5.9000 \end{bmatrix}, \quad
\begin{bmatrix} {}^4 S^5 \\ {}^4 S_O^5 \end{bmatrix} = \begin{bmatrix} -0.5 \\ 0.5 \\ 0.7071 \\ 14.1612 \\ -17.5612 \\ 22.4311 \end{bmatrix}, \quad
\begin{bmatrix} {}^5 S^6 \\ {}^5 S_O^6 \end{bmatrix} = \begin{bmatrix} 0.0795 \\ -0.7866 \\ 0.6124 \\ 23.6061 \\ -10.4425 \\ -16.4759 \end{bmatrix},
$$

where the directions of the lines are dimensionless, and the moments have units of inches. In that previous example, the desired velocity state for the last link of the manipulator with respect to ground was found to be specified as

$$
\begin{bmatrix} {}^0 \omega^6 \\ {}^0 v_O^6 \end{bmatrix} = \begin{bmatrix} 0.200 \\ -0.120 \\ 0.100 \\ 10.6913 \\ 4.7221 \\ -10.9160 \end{bmatrix},
$$

where the angular velocity terms have units of rad/sec, and the linear velocity terms have units of in/sec. The angular velocities of the individual joints that would achieve this velocity state were calculated as

$$
\omega = \begin{bmatrix} -0.0496 \\ 0.9929 \\ -1.6525 \\ -0.0592 \\ 0.4122 \\ -0.2316 \end{bmatrix} \text{rad/sec,} \quad (7.62)
$$

where $\omega = [{}_0\omega_1, {}_1\omega_2, {}_2\omega_3, {}_3\omega_4, {}_4\omega_5, {}_5\omega_6]^T$.

At this instant, the angular accelerations of the individual joints are specified as

$$\alpha = \begin{bmatrix} 1.250 \\ 0.500 \\ -0.750 \\ -1.600 \\ 1.050 \\ -1.900 \end{bmatrix} \text{rad/sec}^2, \tag{7.63}$$

where $\alpha = [_0\alpha_1, {}_1\alpha_2, {}_2\alpha_3, {}_3\alpha_4, {}_4\alpha_5, {}_5\alpha_6]^T$. It is desired to determine the acceleration state parameters of the last link of the manipulator relative to ground, i.e., $^0\alpha^6$ and $^0a_O^6$.

The six twists may be determined from

$$\begin{bmatrix} {}^k\omega^{k+1} \\ {}^kv_O^{k+1} \end{bmatrix} = {}_k\omega_{k+1} \begin{bmatrix} {}^kS^{k+1} \\ {}^kS_O^{k+1} \end{bmatrix}, \quad k = 0\ldots 5 \tag{7.64}$$

and the six twists for this case are calculated as

$$\begin{bmatrix} {}^0\omega^1 \\ {}^0v_O^1 \end{bmatrix} = \begin{bmatrix} 0 \\ 0 \\ -0.0496 \\ 0 \\ 0 \\ 0 \end{bmatrix}, \begin{bmatrix} {}^1\omega^2 \\ {}^1v_O^2 \end{bmatrix} = \begin{bmatrix} -0.7021 \\ 0.7021 \\ 0 \\ 0 \\ 0 \\ 0 \end{bmatrix}, \begin{bmatrix} {}^2\omega^3 \\ {}^2v_O^3 \end{bmatrix} = \begin{bmatrix} 1.1685 \\ -1.1685 \\ 0 \\ 9.9320 \\ 9.9320 \\ -24.3284 \end{bmatrix}, \tag{7.65}$$

$$\begin{bmatrix} {}^3\omega^4 \\ {}^3v_O^4 \end{bmatrix} = \begin{bmatrix} -0.0419 \\ -0.0419 \\ 0 \\ 0.3896 \\ -0.3896 \\ 0.3496 \end{bmatrix}, \begin{bmatrix} {}^4\omega^5 \\ {}^4v_O^5 \end{bmatrix} = \begin{bmatrix} -0.2061 \\ 0.2061 \\ 0.2915 \\ 5.8375 \\ -7.2391 \\ 9.2465 \end{bmatrix}, \begin{bmatrix} {}^5\omega^6 \\ {}^5v_O^6 \end{bmatrix} = \begin{bmatrix} -0.0184 \\ 0.1822 \\ -0.1418 \\ -5.4679 \\ 2.4188 \\ 3.8163 \end{bmatrix},$$

where the angular velocities are expressed in units of rad/sec and the linear velocities in in/sec.

From (7.57) and (7.58), the individual acceleration states of body $k+1$ with respect to body k may be written as

$$^k\alpha^{k+1} = {}_k\alpha_{k+1}\, {}^kS^{k+1}, \tag{7.66}$$

$$^ka_O^{k+1} = {}_k\alpha_{k+1}\, {}^kS_{OL}^{k+1} + {}^k\omega^{k+1} \times {}^kv_O^{k+1}, k = 0\ldots 5. \tag{7.67}$$

The acceleration states are evaluated as

$$
\begin{bmatrix} {}^0\alpha^1 \\ {}^0a_O^1 \end{bmatrix} = \begin{bmatrix} 0 \\ 0 \\ 1.25 \\ 0 \\ 0 \\ 0 \end{bmatrix}, \quad \begin{bmatrix} {}^1\alpha^2 \\ {}^1a_O^2 \end{bmatrix} = \begin{bmatrix} -0.3536 \\ 0.3536 \\ 0 \\ 0 \\ 0 \end{bmatrix}, \quad \begin{bmatrix} {}^2\alpha^3 \\ {}^2a_O^3 \end{bmatrix} = \begin{bmatrix} 0.5303 \\ -0.5303 \\ 0 \\ 32.9349 \\ 32.9349 \\ 12.1688 \end{bmatrix},
$$

$$
\begin{bmatrix} {}^3\alpha^4 \\ {}^3a_O^4 \end{bmatrix} = \begin{bmatrix} -1.1314 \\ -1.1314 \\ 0 \\ 10.5071 \\ -10.5071 \\ 9.4727 \end{bmatrix}, \quad \begin{bmatrix} {}^4\alpha^5 \\ {}^4a_O^5 \end{bmatrix} = \begin{bmatrix} -0.5250 \\ 0.5250 \\ 0.7425 \\ 18.8851 \\ -14.8319 \\ 23.8416 \end{bmatrix}, \quad \begin{bmatrix} {}^5\alpha^6 \\ {}^5a_O^6 \end{bmatrix} = \begin{bmatrix} -0.1510 \\ 1.4945 \\ -1.1635 \\ -43.8131 \\ 20.6865 \\ 32.2560 \end{bmatrix},
$$

$$
\tag{7.68}
$$

where the angular acceleration terms have units of rad/sec^2, and the linear acceleration terms have units of in/sec^2.

Equations (7.41) and (7.42) may now be used to solve for the acceleration state parameters ${}^0\alpha^6$ and ${}^0a_O^6$ as

$$
{}^0\alpha^6 = \begin{bmatrix} -1.7949 \\ 0.6033 \\ -1.1635 \end{bmatrix} \text{ rad/sec}^2 \tag{7.69}
$$

$$
{}^0a_O^6 = \begin{bmatrix} -28.6202 \\ -19.9705 \\ 47.3874 \end{bmatrix} \text{ in/sec}^2.
$$

7.10   Reverse Acceleration Analysis for Serial Manipulators

For the reverse acceleration analysis of a six-axis serial manipulator, the desired acceleration state parameters of the last link relative to ground will be specified, i.e., ${}^0\alpha^6$ and ${}^0a_O^6$. Additionally, the individual joint angles, ${}_k\theta_{k+1}$, (or joint, offsets for a prismatic joint, ${}_kS_{k-1}$) and the individual joint angular velocities, ${}_k\omega_{k+1}$ (or linear velocity for a prismatic joint, ${}_kv_{k+1}$), $k = 0\ldots5$, are assumed known, in addition to the constant mechanism parameters. The objective is to determine the magnitudes of the relative joint angular accelerations, ${}_k\alpha_{k+1}$ (or linear accelerations ${}_ka_{k+1}$ for a prismatic joint), $k = 0\ldots5$ that will achieve the desired acceleration state for the last link.

For the case where all six joints are screw joints (or revolute joints with zero pitch), substituting (7.43) into (7.41) and rearranging gives

$$
{}^0\alpha_R^6 = {}_0\alpha_1 \, {}^0S^1 + {}_1\alpha_2 \, {}^1S^2 + {}_2\alpha_3 \, {}^2S^3 + {}_3\alpha_4 \, {}^3S^4 + {}_4\alpha_5 \, {}^4S^5 + {}_5\alpha_6 \, {}^5S^6, \tag{7.70}
$$

where

$$^0\alpha_R^6 = {}^0\alpha^6 - {}^0\omega^1 \times \left({}^1\omega^2 + {}^2\omega^3 + {}^3\omega^4 + {}^4\omega^5 + {}^5\omega^6\right) \qquad (7.71)$$
$$- {}^1\omega^2 \times ({}^2\omega^3 + {}^3\omega^4 + {}^4\omega^5 + {}^5\omega^6)$$
$$- {}^2\omega^3 \times ({}^3\omega^4 + {}^4\omega^5 + {}^5\omega^6)$$
$$- {}^3\omega^4 \times ({}^4\omega^5 + {}^5\omega^6)$$
$$- {}^4\omega^5 \times ({}^5\omega^6).$$

Similarly, substituting (7.56) into (7.42) and rearranging gives

$$^0a_{OR}^6 = {}_0\alpha_1 \, {}^0S_O^1 + {}_1\alpha_2 \, {}^1S_O^2 + {}_2\alpha_3 \, {}^2S_O^3 + {}_3\alpha_4 \, {}^3S_O^4 + {}_4\alpha_5 \, {}^4S_O^5 + {}_5\alpha_6 \, {}^5S_O^6, \qquad (7.72)$$

where

$$^0a_{OR}^6 = {}^0a_O^6 - \left({}^0\omega^1 \times {}^0v_O^1\right) - \left({}^1\omega^2 \times {}^1v_O^2\right) - \left({}^2\omega^3 \times {}^2v_O^3\right) \qquad (7.73)$$
$$- \left({}^3\omega^4 \times {}^3v_O^4\right) - \left({}^4\omega^5 \times {}^4v_O^5\right) - \left({}^5\omega^6 \times {}^5v_O^6\right)$$
$$- 2{}^0\omega^1 \times \left({}^1v_O^2 + {}^2v_O^3 + {}^3v_O^4 + {}^4v_O^5 + {}^5v_O^6\right)$$
$$- 2{}^1\omega^2 \times \left({}^2v_O^3 + {}^3v_O^4 + {}^4v_O^5 + {}^5v_O^6\right)$$
$$- 2{}^2\omega^3 \times \left({}^3v_O^4 + {}^4v_O^5 + {}^5v_O^6\right)$$
$$- 2{}^3\omega^4 \times \left({}^4v_O^5 + {}^5v_O^6\right)$$
$$- 2{}^4\omega^5 \times \left({}^5v_O^6\right).$$

As was the case for the forward acceleration analysis, since all the joint angles are known, the Plücker line coordinates of all the joint axes of the manipulator can be determined, i.e., $\{{}^kS^{k+1}; {}^kS_O^{k+1}\}$, $k = 0\ldots 5$. Further, the coordinates of each of the six screws, $\{{}^kS^{k+1}; {}^kS_O^{k+1}\}$, $k = 0\ldots 5$, can be determined by incorporating the pitch term of each axis. Since the joint angular velocities are known, the instantaneous twist of body k relative to body $k + 1$ may be obtained as $\{{}^k\omega^{k+1}; {}^kv_O^{k+1}\} = {}_k\omega_{k+1}\{{}^kS^{k+1}; {}^kS_O^{k+1}\}$, $k = 0\ldots 5$. At this point, the terms ${}^0\alpha_R^6$ and ${}^0a_{OR}^6$, as defined in (7.71) and (7.73), are determined. Equations (7.70) and (7.72) may be written in matrix format as

$$\begin{bmatrix} {}^0\alpha_R^6 \\ {}^0a_{OR}^6 \end{bmatrix} = J \begin{bmatrix} {}_0\alpha_1 \\ {}_1\alpha_2 \\ {}_2\alpha_3 \\ {}_3\alpha_4 \\ {}_4\alpha_5 \\ {}_5\alpha_6 \end{bmatrix}, \qquad (7.74)$$

where J is a 6×6 matrix whose columns are the coordinates of the six screws of the manipulator. J is written as

$$J = \begin{bmatrix} {}^0S^1 & {}^1S^2 & {}^2S^3 & {}^3S^4 & {}^4S^5 & {}^5S^6 \\ {}^0S_O^1 & {}^1S_O^2 & {}^2S_O^3 & {}^3S_O^4 & {}^4S_O^5 & {}^5S_O^6 \end{bmatrix}. \qquad (7.75)$$

The individual joint accelerations may be obtained as

$$
\begin{bmatrix} {}_0\alpha_1 \\ {}_1\alpha_2 \\ {}_2\alpha_3 \\ {}_3\alpha_4 \\ {}_4\alpha_5 \\ {}_5\alpha_6 \end{bmatrix} = \boldsymbol{J}^{-1} \begin{bmatrix} {}^0\boldsymbol{\alpha}_R^6 \\ {}^0\boldsymbol{a}_{OR}^6 \end{bmatrix}. \tag{7.76}
$$

It is interesting to note that the reverse acceleration problem cannot be solved if the matrix $\boldsymbol{J}$ is singular, which would occur if the set of screws were linearly dependent. Since the screw coordinates are a function of the joint positions, there may be configurations of the manipulator where $\boldsymbol{J}$ is singular. This is the same condition whereby it was not possible to solve the reverse velocity problem for the serial manipulator.

7.10.1   Sample Problem

Consider the numerical example in the previous section. For the same position and velocity conditions, it is now desired to determine the angular accelerations of the individual joints of the Puma manipulator so that the acceleration state parameters of the last link of the manipulator relative to ground equal some desired values. For this case, the joint angular accelerations will be determined such that the angular acceleration of the last link equals zero, i.e., ${}^0\boldsymbol{\alpha}^6 = [0,0,0]^T$, and the point in the last link coincident with the origin will be accelerating in the z axis direction at a rate of 2.5 in/sec^2, i.e., ${}^0\boldsymbol{a}_O^6 = [0,0,2.5]^T$ in/sec^2.

The velocity state parameters for the individual joints were listed in (7.65). Equation (7.71) can be used to calculate ${}^0\boldsymbol{\alpha}_R^6$ as

$$
{}^0\boldsymbol{\alpha}_R^6 = \begin{bmatrix} 0.1644 \\ 0.1081 \\ 0.0221 \end{bmatrix} \text{ rad/sec}^2 \tag{7.77}
$$

and (7.73) can be used to calculate ${}^0\boldsymbol{a}_{OR}^6$ as

$$
{}^0\boldsymbol{a}_{OR}^6 = \begin{bmatrix} 13.6675 \\ 15.3580 \\ 8.3678 \end{bmatrix} \text{ in/sec}^2. \tag{7.78}
$$

The matrix $\boldsymbol{J}$ can be evaluated by substituting (7.61) into (7.75). Finally, the joint angular accelerations can be obtained from (7.76) as

$$
\begin{bmatrix} {}_0\alpha_1 \\ {}_1\alpha_2 \\ {}_2\alpha_3 \\ {}_3\alpha_4 \\ {}_4\alpha_5 \\ {}_5\alpha_6 \end{bmatrix} = \boldsymbol{J}^{-1} \begin{bmatrix} {}^0\boldsymbol{\alpha}_R^6 \\ {}^0\boldsymbol{a}_{OR}^6 \end{bmatrix} = \begin{bmatrix} 0.0033 \\ 2.6322 \\ -5.5450 \\ -0.9726 \\ 2.0449 \\ -2.3305 \end{bmatrix} \text{ rad/sec}^2. \tag{7.79}
$$

7.11   Problems

1. A planar mechanism is shown in Figure 7.8. At the instant shown, the coordinates of points A and B (given in units of inches) are:

$$r_A = 5.75i + 10j$$
$$r_B = 15i + 10j.$$

The angular velocity of body 1 with respect to body 0 is given as $_0\omega_1 = 2$ rad/sec in the direction shown. The angular acceleration of link 1 with respect to link 0 is 0.5 rad/sec², and it is in the same direction as $_0\omega_1$. Determine the angular velocity and angular acceleration of body 3 with respect to body 0 (magnitude and direction). Also, determine the linear velocity and linear acceleration of point A in body 2 with respect to body 0.

2. A velocity analysis of the mechanism shown in Figure 7.9 was performed in Problem 4 in Chapter 4. At the instant shown in the figure, it is desired that the end effector translate at a speed of 4 cm/sec in a direction parallel to the vector $3i + 4j$. From the previous problem, the coordinates of the screws along the three twists were determined as

$$
\begin{bmatrix} ^0S^1 \\ ^0S_O^1 \end{bmatrix} = \begin{bmatrix} 0 \\ 0 \\ 1 \\ 0 \\ 0 \\ 0 \end{bmatrix}, \quad
\begin{bmatrix} ^1S^2 \\ ^1S_O^2 \end{bmatrix} = \begin{bmatrix} 0 \\ 0 \\ 0 \\ 1 \\ 0 \\ 0 \end{bmatrix}, \quad
\begin{bmatrix} ^2S^3 \\ ^2S_O^3 \end{bmatrix} = \begin{bmatrix} 0 \\ 0 \\ 1 \\ 5 \\ -4 \\ 0 \end{bmatrix}.
$$

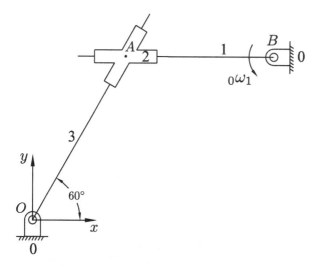

Figure 7.8 Planar mechanism

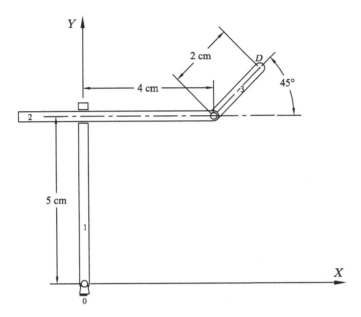

Figure 7.9 RPR manipulator

Values for the joint parameter velocities $_0\omega_1$, $_1v_2$, and $_2\omega_3$ that caused the end effector to translate as desired were determined to be

$$_0\omega_1 = 0.8 \text{ rad/sec}, \quad _1v_2 = 6.4 \text{ cm/sec}, \quad _0\omega_1 - 0.8 \text{ rad/sec}.$$

At the instant shown, it is desired that the linear acceleration of point D be equal to 5 cm/sec² in a direction parallel to the Y axis. Also, the angular acceleration of body 3 with respect to ground (body 0) is desired to be equal to $10\,\boldsymbol{k}$ rad/sec². Determine the following:

- Desired acceleration state of body 3 with respect to ground in terms of the coordinate system shown in the figure.
- Angular accelerations about the first and third joint axes and linear acceleration about the second joint axis that will generate the desired acceleration state, i.e., determine values for the scalars $_0\alpha_1$, $_1a_2$, and $_2\alpha_3$.

3. A manipulator is shown in Figure 7.10. Linear dimensions are shown in meters. At the instant shown, it is desired that the last link be moving in pure translation in the Y direction at a speed of 2.5 m/sec. Further, it is desired that the last link be accelerating in the Y direction at a rate of 1.5 m/sec² (with no angular acceleration). Determine values for the required joint velocities and accelerations that will accomplish this.

4. In Problem 3 in Chapter 4, the angular velocities for the Cincinnati Milacron T3-776 robot manipulator were calculated in order to achieve a desired velocity state for the end effector. For the same position and velocity state of the manipulator, determine the angular accelerations of each of the six joints such that

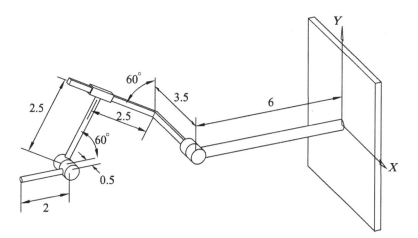

Figure 7.10 RPR manipulator

the angular acceleration of the end effector with respect to ground will be $^0\boldsymbol{\alpha}^6 = [2, 0.5, 0]^T$ rad/sec^2 and the linear acceleration of the point in the end effector that is coincident with the origin is $^0\boldsymbol{a}^6_O = [20, 0, -10]^T$ in/sec^2.

5. In Problem 5 of Chapter 4, the velocity of body 1 was given as $[0, -6, 0]^T$ in/sec, and the velocity of body 4 was given as $[4, 0, 0]^T$ in/sec. At the instant shown in Figure 4.22, body 1 is accelerating upward at a rate of 5 in/sec^2, and body 4 is accelerating to the right at a rate of 6 in/sec^2 with respect to body 0. Determine the acceleration of point D measured with respect to body 0 at this instant.

Appendix A Derivation of Cylindroid Equations

In Section 3.9, two wrenches were considered to be acting on a rigid body. For convenience, and without loss of generality, the reference point O was chosen at the intersection of the line of action of the first wrench and the mutually perpendicular line between the two wrenches (see Figure 3.13). The perpendicular distance between the lines of action of these wrenches is a_{12}, and a_{12} is a unit vector along the line that is mutually perpendicular to the lines of action of the two wrenches. The angle between f_1 and f_2 is labeled as α_{12} and is measured in a right hand sense about a_{12}. The coordinates of the two given wrenches are, thus, written as $\{f_1; h_1 f_1\}$ and $\{f_2; a_{12}a_{12} \times f_2 + h_2 f_2\}$.

For the two given wrenches, the values for the magnitude, f, orientation angle, ψ, pitch, h, and location, r, for the resultant of the two original wrenches can be determined from equations (3.78), (3.79) and (3.81), (3.88), and (3.92), respectively. It was pointed out in Section 3.10 that, as the ratio of the magnitudes of the applied screws, $\frac{f_1}{f_2}$, varies, the orientation angle of the resultant, ψ, varies between 0 and 2π. Equations (3.95) and (3.96) define the location of the resultant, r, and the pitch of the resultant, h, in terms of the resultant orientation angle, ψ. These equations will now be derived.

The pitch of the resultant screw was expressed in equation (3.88) and is repeated here as

$$h = \frac{f_1}{f} h_1 \cos \psi + \frac{f_2}{f} h_2 \cos(\alpha_{12} - \psi) - \frac{f_1 f_2}{f^2} a_{12} \sin \alpha_{12}. \tag{A.1}$$

Subtracting h_1 from both sides and rearranging yields

$$h - h_1 = h_1 \left(\frac{f_1}{f} \cos \psi - 1 \right) + h_2 \left(\frac{f_2}{f} \cos(\alpha_{12} - \psi) \right) - \frac{f_1 f_2}{f^2} a_{12} \sin \alpha_{12}, \tag{A.2}$$

which can be written as

$$h - h_1 = A h_1 + B h_2 + D a_{12}, \tag{A.3}$$

where

$$A = \frac{f_1}{f} \cos \psi - 1, \tag{A.4}$$

$$B = \frac{f_2}{f} \cos(\alpha_{12} - \psi), \tag{A.5}$$

$$D = -\frac{f_1 f_2}{f^2} \sin \alpha_{12}. \tag{A.6}$$

From (3.80), the ratio $\frac{f_1}{f}$ may be written as

$$\frac{f_1}{f} = \frac{\sin(\alpha_{12} - \psi)}{\sin \alpha_{12}}. \tag{A.7}$$

Expanding the numerator as $\sin(\alpha_{12} - \psi) = \sin \alpha_{12} \cos \psi - \cos \alpha_{12} \sin \psi$ yields

$$\frac{f_1}{f} = \cos \psi - \cot \alpha_{12} \sin \psi. \tag{A.8}$$

From (3.79), the ratio $\frac{f_2}{f}$ may be written as

$$\frac{f_2}{f} = \frac{\sin \psi}{\sin \alpha_{12}}. \tag{A.9}$$

Substituting (A.8) into (A.4) yields

$$A = \cos^2 \psi - \cot \alpha_{12} \sin \psi \cos \psi - 1. \tag{A.10}$$

Substituting $-\sin^2 \psi = \cos^2 \psi - 1$ gives

$$A = -\left(\sin^2 \psi + \cot \alpha_{12} \sin \psi \cos \psi \right). \tag{A.11}$$

Introducing the trigonometric identities $\sin 2\psi = 2 \sin \psi \cos \psi$ and $\cos 2\psi = 1 - 2 \sin^2 \psi$ gives

$$A = \frac{1}{2} \left(\cos 2\psi - \cot \alpha_{12} \sin 2\psi - 1 \right). \tag{A.12}$$

Substituting (A.9) into (A.5) yields

$$B = \frac{\sin \psi}{\sin \alpha_{12}} \cos(\alpha_{12} - \psi). \tag{A.13}$$

Substituting $\cos(\alpha_{12} - \psi) = \cos \alpha_{12} \cos \psi + \sin \alpha_{12} \sin \psi$ and rearranging gives

$$B = \sin \psi \cos \psi \cot \alpha_{12} + \sin^2 \psi. \tag{A.14}$$

Upon comparing (A.14) and (A.11) it is apparent that $B = -A$.
Substituting (A.8) and (A.9) into (A.6) yields

$$D = -\frac{\sin \psi}{\sin \alpha_{12}} \left(\cos \psi - \cot \alpha_{12} \sin \psi \right) \sin \alpha_{12}. \tag{A.15}$$

Simplifying this equation gives

$$D = -\sin \psi (\cos \psi - \cot \alpha_{12} \sin \psi), \tag{A.16}$$

$$D = -\sin \psi \cos \psi + \cot \alpha_{12} \sin^2 \psi. \tag{A.17}$$

Introducing the trigonometric identities $\sin 2\psi = 2 \sin \psi \cos \psi$ and $\cos 2\psi = 1 - 2 \sin^2 \psi$ gives

$$D = -\frac{1}{2} \left(\sin 2\psi + \cot \alpha_{12}(\cos 2\psi - 1) \right). \tag{A.18}$$

Substituting (A.12), $B = -A$, and (A.18) into (A.3) and regrouping terms gives

$$h - h_1 = -\frac{1}{2} \left(\cos 2\psi - \cot \alpha_{12} \sin 2\psi - 1 \right) (h_2 - h_1) \tag{A.19}$$
$$- \frac{1}{2} \left(\sin 2\psi + \cot \alpha_{12}(\cos 2\psi - 1) \right) a_{12}.$$

Rearranging this equation yields

$$h - h_1 = \frac{1}{2} \left[(h_2 - h_1) + a_{12} \cot \alpha_{12} \right] \tag{A.20}$$
$$- \frac{1}{2} \left[(h_2 - h_1) + a_{12} \cot \alpha_{12} \right] \cos 2\psi$$
$$- \frac{1}{2} \left[a_{12} - (h_2 - h_1) \cot \alpha_{12} \right] \sin 2\psi.$$

The term R, which will have units of length, is now defined as

$$R = \sqrt{\frac{a_{12}^2 + (h_2 - h_1)^2}{4 \sin^2 \alpha_{12}}}. \tag{A.21}$$

In addition, the sine and cosine of an angle σ are defined as

$$\sin \sigma = \frac{a_{12} - (h_2 - h_1) \cot \alpha_{12}}{2R}, \tag{A.22}$$

$$\cos \sigma = \frac{(h_2 - h_1) + a_{12} \cot \alpha_{12}}{2R}. \tag{A.23}$$

Note that $\sin^2 \sigma + \cos^2 \sigma = 1$ and that R, $\sin \sigma$, and $\cos \sigma$ are all defined in terms of known quantities, i.e., the pitches of the two given screws, the angle between the two given screws, and the perpendicular distance between the two given screws.

Multiplying (A.22) and (A.23) by $2R$ and substituting into (A.20) yields

$$h - h_1 = \frac{1}{2}(2R \cos \sigma) - \frac{1}{2}(2R \cos \sigma) \cos 2\psi - \frac{1}{2}(2R \sin \sigma) \sin 2\psi. \tag{A.24}$$

Expanding this equation gives

$$h - h_1 = R \cos \sigma (1 - \cos 2\psi) - R \sin \sigma \sin 2\psi$$
$$= R(\cos \sigma - \cos \sigma \cos 2\psi - \sin \sigma \sin 2\psi)$$
$$= R(\cos \sigma - \cos(2\psi - \sigma)). \tag{A.25}$$

Equation (A.25) is presented in Section 3.10 as equation (3.96).

The position of the resultant screw was expressed in equation (3.92) and is repeated here as

$$r \sin \psi = \frac{f_1}{f}(h - h_1) + \frac{f_2}{f}\{(h - h_2) \cos \alpha_{12} + a_{12} \sin \alpha_{12}\}. \tag{A.26}$$

Dividing throughout by $\sin \psi$ and substituting (A.8) and (A.9) gives

$$r = \frac{1}{\sin \psi}[(\cos \psi - \cot \alpha_{12} \sin \psi)(h - h_1) \tag{A.27}$$

$$+ \frac{\sin \psi}{\sin \alpha_{12}} \{(h - h_2) \cos \alpha_{12} + a_{12} \sin \alpha_{12}\}].$$

Simplifying this expression gives

$$r = (\cot \psi - \cot \alpha_{12})(h - h_1) + (h - h_2) \cot \alpha_{12} + a_{12} \tag{A.28}$$

$$= \cot \psi(h - h_1) - (h_2 - h_1) \cot \alpha_{12} + a_{12}.$$

The term $(h - h_1)$ will be replaced by substituting (A.11) and (A.17) together with the fact that $B = -A$ into (A.3) to yield

$$h - h_1 = \left(\sin^2 \psi + \cot \alpha_{12} \sin \psi \cos \psi\right)(h_2 - h_1) \tag{A.29}$$

$$+ a_{12}(- \sin \psi \cos \psi + \cot \alpha_{12} \sin^2 \psi).$$

Multiplying throughout by $\cot \psi$ yields

$$\cot \psi(h - h_1) = (\sin \psi \cos \psi + \cot \alpha_{12} \cos^2 \psi)(h_2 - h_1) \tag{A.30}$$

$$+ a_{12}(- \cos^2 \psi + \cot \alpha_{12} \sin \psi \cos \psi).$$

Substituting $\sin 2\psi = 2 \sin \psi \cos \psi$ and $\cos 2\psi = 2 \cos^2 \psi - 1$ gives

$$\cot \psi(h - h_1) = \frac{1}{2}(\sin 2\psi + \cot \alpha_{12}(\cos 2\psi + 1))(h_2 - h_1) \tag{A.31}$$

$$- \frac{1}{2}a_{12}(\cos 2\psi + 1 - \cot \alpha_{12} \sin 2\psi).$$

Substituting (A.31) in (A.28) yields

$$r = \frac{1}{2}(\sin 2\psi + \cot \alpha_{12}(\cos 2\psi + 1))(h_2 - h_1) \tag{A.32}$$

$$- \frac{1}{2}a_{12}(\cos 2\psi + 1 - \cot \alpha_{12} \sin 2\psi) - (h_2 - h_1) \cot \alpha_{12} + a_{12}.$$

Regrouping this equation gives

$$r = \frac{1}{2}[(h_2 - h_1) + a_{12} \cot \alpha_{12}] \sin 2\psi \tag{A.33}$$

$$+ \frac{1}{2}[(h_2 - h_1) \cot \alpha_{12} - a_{12}] \cos 2\psi + \frac{1}{2}[a_{12} - (h_2 - h_1) \cot \alpha_{12}].$$

Substituting (A.22) and (A.23) gives

$$r = [R \cos \sigma] \sin 2\psi + [-R \sin \sigma] \cos 2\psi + [R \sin \sigma] \qquad \text{(A.34)}$$
$$= R(\sin \sigma + \sin(2\psi - \sigma)).$$

Equation (A.34) expresses the location of the resultant wrench, r, as a function of the resultant orientation angle ψ and is presented in Section 3.10 as equation (3.95).

Bibliography

Ball, Sir R. S. (1900). *A Treatise on the Theory of Screws*. (Reprinted: 1998). Cambridge University Press.

Brand, Louis. (1947). *Vector and Tensor Analysis*. (Newer reprints have fixed some typos). Wiley.

Coxeter, H. S. M. (2003). *Projective Geometry*. (2nd ed., originally 1974). Springer.

Crane, Carl D. III, and Joseph Duffy. (2008). *Kinematic Analysis of Robot Manipulators*. Cambridge University Press.

Davidson, Joseph K., and Kenneth H. Hunt. (2004). *Robots and Screw Theory: Applications of Kinematics and Statics to Robotics*. Oxford University Press.

Dimentberg, F. M. (1968). *The Screw Calculus and Its Applications in Mechanics*. (Document AD680993. Translated from Russian 1965 version). US Dept. of Commerce National Technical Information Service.

Duffy, Joseph. (1980). *Analysis of Mechanisms and Robot Manipulators*. Wiley.

Faulkner, T. Ewan. (2006). *Projective Geometry*. (1960 edition). Dover.

Grassmann, Hermann. (1862). *Extension Theory*. (Translated in 2000 by Lloyd C. Kannenberg from the German 1862 version entitled Ausdehnungslehre). American Mathematical Society.

Hartenberg, Richard S., and Jacques Denavit. (1964). *Kinematic Synthesis of Linkages*. McGraw-Hill.

Hunt, Kenneth H. (1978). *Kinematic Geometry of Mechanisms*. Clarendon Press.

——— (1983). "Structural kinematics of in-parallel-actuated robots arms." *ASME Journal of Mechanisms, Transmissions, and Automation in Design* 105 (4): 705–712.

Klein, Felix. (1939). *Elementary Mathematics from an Advanced Standpoint: Geometry*. (Book was translated from German 3ed in 1939 by Hedrick and Noble. Klein was a student of Plücker). Dover.

Meserve, Bruce E. (2010). *Fundamental Concepts of Geometry*. Dover.

Plücker, Julius. (1865). "On a new geometry of space." (Plücker coordinates discussed near end of article). *Philosophical Transactions of the Royal Society of London* 155: 725–791.

——— (1866). "Fundamental views regarding mechanics." (Plücker Dyname article.) *Philosophical Transactions of the Royal Society of London* 156: 361–380.

Rico, J. M., J. Gallardo, and B. Ravani. (2003). "Lie algebra and the mobility of kinematic chains." (Joseph Duffy Memorial Edition), *Journal of Robotic Systems* 20 (8): 477–499.

Scott, Charlotte Angas. (1894). *An Introductory Account of Certain Modern Ideas and Methods in Plane Analytical Geometry*. (Scott was a student of Cayley). Macmillan.

Selig, J. M. (2005). *Geometric Fundamentals of Robotics*. Springer-Verlag New York.

Semple, J. G., and G. T. Kneebone. (1998). *Algebraic Projective Geometry*. Oxford.

Sugimoto, Koichi, and Joseph Duffy. (1982). "Application of linear algebra to screw systems." *Mechanism and Machine Theory* 17 (1): 73–83.

Index

Printed in the United States
by Baker & Taylor Publisher Services